Statistics in Geography and Environmental Science

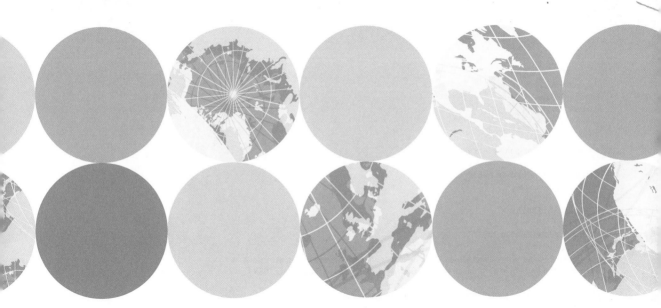

Statistics in Geography and Environmental Science

Richard Harris
University of Bristol

Claire Jarvis
University of Leicester

Prentice Hall
is an imprint of

Harlow, England • London • New York • Boston • San Francisco • Toronto
Sydney • Tokyo • Singapore • Hong Kong • Seoul • Taipei • New Delhi
Cape Town • Madrid • Mexico City • Amsterdam • Munich • Paris • Milan

Pearson Education Limited
Edinburgh Gate
Harlow
Essex CM20 2JE
England

and Associated Companies throughout the world

Visit us on the World Wide Web at:
www.pearsoned.co.uk

First published 2011

ISBN: 978-0-13-178933-3

British Library Cataloguing-in-Publication Data
A catalogue record for this book is available from the British Library

Library of Congress Cataloging-in-Publication Data
Harris, Richard, 1973-
    Statistics in geography and environmental science/Richard Harris, Claire Jarvis.
        p. cm.
    Includes bibliographical references and index.
    ISBN 978-0-13-178933-3 (pbk.: alk. paper)
    1. Geography—Statistical methods.   2. Environmental sciences—Statistical methods.
I. Jarvis, Claire.   II. Title.
    G70.3.H37 2011
    910.01'5195—dc22

                                                        2011007382

10 9 8 7 6 5 4 3 2 1
14 13 12 11

Typeset in 10.5/13pt Minion Pro by 75
Printed and bound by Ashford Colour Press Ltd, Gosport

Brief contents

Contents

Preface

There are plenty of books about geography and statistics. It is unlikely the world is crying out for another. Why, then, seek to reinvent the wheel?

Well, wheels are being reinvented all the time. The wheels on our family-friendly saloons are different from those on the Model T Ford; technologies change, so too do the road conditions and the journeys from where we have been to where we are heading.

There are many excellent textbooks covering statistics in geography that we happily recommend. Our aim in writing this book was not to compete. Quite the opposite. We hope that after reading this text you will be encouraged and equipped to read the others, taking your knowledge and understanding further as you do so.

We imagine our audience to be an undergraduate class either taking statistics for the first time or taking some of the more basic ideas further. Because we believe in learning by doing, this book is intended to support the sort of applied project work an instructor is likely to set. However, if you aim to be self-taught, this book will assist you, too.

It would take a determined enthusiast (and most likely a masochist) to read this book in a single sitting. Although the book does proceed in a cumulative manner, with later chapters building on earlier ones, we have helped you to dip in and out of the book, and to find what you are looking for, by providing chapter summaries, learning objectives and summaries of key statistical concepts that can be referred back to without reading the whole chapter again.

It has been important to us to focus on concepts and ideas, showing how they are expressed in equations and statistical ways of working. Many people tend to be discouraged by increasingly elaborate equations and it is possible to write a textbook without them. We did not. Our preference is to confront them head on: to demystify the notation and unpack what it is saying (and assuming).

Statistics and numbers convey a sense of authority, a statement of fact, that this is the right answer, beyond challenge. That is nonsense, of course. Statistics are simply a tool for social and scientific knowledge forming and debate – a useful tool but not a perfect one. Knowing when and how to use them, and what is presumed by doing so, is a useful transferable skill whether your goal is to enter a job requiring numeracy or you wish to challenge the way 'management by numbers' and a statistical spin now seem to permeate all areas of social life and governance.

The book does not pretend to be a treatise in statistics in the sense of beginning from the first principles of probability theory and working through the various types of distributions that emerge under particular conditions and constraints. It is very much a handbook for applied data analysis; all the examples we give are based on analyses of real data sets. We do, however, proceed in the broadly conventional way from descriptive statistics, though inferential statistics to relational statistics.

Though much of the theory and thinking in this book is many years old (greatly predating the authors) we have taken the opportunity to bring things up to date where it is appropriate to do so. This is especially true in the final chapter where we discuss spatial regression models, geographically weighted regression and multilevel modelling, albeit in an introductory way.

We have also made ample use of developments in open source (statistical) computing. Though the textbook is not tied to any particular software – it introduces the concepts, not how they are deployed in one of many statistical packages – it is no secret that most of the analyses and graphics were produced in R (www.r-project.org). We have placed a high value on the use of graphics, including maps, in this book. Used effectively, they are essential for making sense of data, exploring them, avoiding silly errors and supporting the knowledge-forming process. Used dishonestly, they are an effective way of lying.

This book has had a long gestation period, much longer than either of us intended. Throughout this period we have been grateful for the support of our families, colleagues and our commissioning editors, Rufus and Andrew. We are also indebted to Dave Unwin and Chris Keylock for providing helpful and insightful feedback on early drafts of the chapters. We may not have agreed with them on everything, and undoubtedly there are some things they would have presented differently. Nevertheless, they did exactly what a decent referee should do: give good honest advice with an insistence on academic rigour. Thank you.

Appreciation is also due to our colleagues at Bristol and Leicester. RH especially thanks members of the spatial modelling group, past and present: Paul, Ron, Kelvyn, Tony, Malcolm, Winnie and David. Some of the contents of this book were written during a Spatial Literacy in Teaching (SPLINT) fellowship and draw on research funded by the ESRC under grant RES-149-25-1041.

Readers are encouraged to visit www.social-statistics.org where the role of statistics in society is discussed, and where resources to support the teaching and learning of the methods discussed in this book will be posted.

Finally, this book is written in memory of Professor Les Hepple, who taught statistics (amongst other things) to generations of students at the School of Geographical Sciences, University of Bristol. Les was an exceptional scholar and teacher; the clarity of his lectures was unsurpassed. It is a tribute to him that his lecture material continues to be used within the School's teaching. Some wheels are not worth reinventing; they cannot be improved.

Rich Harris, University of Bristol
Claire Jarvis, University of Leicester
August 2010

About this book

There are decisions that need to be taken when writing a textbook such as this. Some of these concern the style and content of the chapters. We are not professional statisticians and this is not a book for statistics students. We are academic geographers who use data, geographical information technologies and statistical methods in some areas of our teaching and research. Consequently, this book is not a treatise on statistics in the manner of Frank and Althoen (1994). **It is a book focused on analysing, exploring and making sense of data in areas of substantive interest to physical, environmental and human geographers, and to geographical information scientists.** Our approach is influenced by Erickson and Nosanchuk (1992) and by Rogerson (2006).

The chapters tend to be data led, meaning we take a data set and analyse it. The reasons for this are largely pedagogic, to demonstrate the concepts we are writing about. It may suggest a preference for research that is data or technique driven, which starts with the data and finds an answer. This is not intended and we warn again about placing too great an emphasis on data collection and processing, and too little on a guiding socio-economic or scientific theory. We recommend reading Plummer and Taylor (2001a; 2001b) as an example of how theories are first formed, *then* modelled and tested using data.

Throughout the book we have assumed our readers have mathematical training to a mid-level at high school. We are conscious of the Stephen Hawking quote 'that each equation I included in the book would halve the sales'. Nevertheless, it seems unwise (and somewhat incongruous) to deny that statistics involve mathematics and equations. Even if those equations operate behind the button of most software packages, it remains a good idea to know how the calculations are made and how their answers are obtained. Equations are therefore included and explained as fully as possible.

By contrast, we have not included any comprehension exercises at the end of each chapter. Our experience is that such exercises are well intended but rather contrived and uninspiring. Instead, we encourage you to get your 'hands dirty' with a real data set, analyse it and see what you learn. For many readers, in-class exercises and instructor-led tutorials will provide exactly the sort of formative experience and project-based learning that we support.

Finally, there are a large number of computer packages that have statistical capability and could be used to accompany this book. Some functions are built into spreadsheet software such as Microsoft's Excel, Sun's StarOffice Calc and OpenOffice's 'freeware' program (www.openoffice.org). They are also available in explicitly statistical software such as SPSS, SAS, Stata and Minitab.

Most of the statistical methods described in this book will be available in these various packages. Yet, every package is different: they do not look the same, do not work the same way and each has its strengths and merits. Which you use is a matter of choice, preference, experience, what is available, and of the research you are undertaking. It may well be determined for you by your class instructor or by the software licences your institution holds.

Although we do not endorse any particular package, we will, as the book progresses, identify more specialist software capable of explicitly geographical analyses, and from where they are obtained (often at no cost). We also draw your attention to R, a free software environment for statistical computing and graphics. It is not the most obvious choice for beginners because it runs using command lines rather than drop-down menus or tabs. However, it is extremely powerful, flexible and supported by a large user community that has developed a whole suite of analytical extras, including mapping and explicitly geographical analysis. Most of the graphics in this book were produced in R. You can download it from www.r-project.org. Introductory texts using R include Verzani (2004) and Crawley (2005), and there are other guides to its use at http://cran.r-project.org/manuals.html. The Use R book series published by Springer includes the books *Applied Spatial Data Analysis with R* (Bivand *et al.* 2008), *Applied Econometrics with R* (Kleiber and Zeileis 2008) and *Statistical Methods for Environmental Epidemiology with R* (Peng and Dominici 2008).

Because we cannot know which is best or available for you, we have decided not to tie this book to any specific piece of software. In any case, we are primarily interested in introducing general theory and concepts. Mixing those with the commands and computing knowledge required for a particular statistical package risks that neither would be understood. The advantage of our approach is clarity. The disadvantage is you will still need to familiarise yourself with the workings of your chosen software.

The book proceeds as follows. In Chapter 2 we look at the sorts of statistics that are in everyday use. They are descriptive and aim to summarise the character and shape of a data set. They include measures of average and of spread. In Chapter 3 we look further at the shape of data sets and introduce one of the more commonly found – the normal curve. The properties of this curve are important, not only because they can be used to detect unusual observations in a batch of data, but also because they provide the foundation upon which many inferential statistics are built. In Chapter 4 we focus on research design and sampling strategies. These are too easily overlooked or brushed over, but they are critical to the success of an undergraduate research project, for example. The bottom line is that if the data collection is unreliable then the results of the analysis will be too.

In Chapter 5, we begin to switch from merely describing a data set in its own right to using it to make inferences (statements) about what it is the data are a sample of. This continues into Chapter 6 where we introduce formal methods of hypothesis testing, investigating, for example, whether two sets of data could, despite their differences, be measurements of the same thing.

Chapter 7 introduces methods to test for relationships but in ways that are essentially blind to geography. Chapter 8 takes a different tact – introducing ways of looking for patterns of spatial association. Chapter 9 bridges the two, extending the relational statistics to accommodate the spatial properties of geographical data. The book ends with an epilogue offering some conclusions, reflections and ideas for further study.

References

Bivand, R.S., Pebesma, E.J. and Gomez-Rubio, V. (2008) *Applied Spatial Data Analysis with R*, New York: Springer.

Crawley, M.J. (2005) *Statistics: An Introduction using R,* Chichester: Wiley.

Erickson, B.H. and Nosanchuk, T.A. (1992) *Understanding Data,* 2nd edn., Maidenhead: Open University Press.

Frank, H. and Althoen, S.C. (1994) *Statistics: Concepts and Applications Workbook,* Cambridge: Cambridge University Press.

Kleiber, C. and Zeileis, A. (2008) *Applied Econometrics with R,* New York: Springer.

Peng, R.D. and Dominici, F. (2008) *Statistical Methods for Environmental Epidemiology with R: A Case Study in Air Pollution and Health,* New York: Springer.

Plummer, P. and Taylor, M. (2001a) Theories of local economic growth (part 1): concepts, models, and measurement. Environment and Planning A, 33(2), 219–236.

Plummer, P. and Taylor, M. (2001b) Theories of local economic growth (part 2): model specification and empirical validation. *Environment and Planning A,* 33(3), 385–398.

Rogerson, P.A. (2006) *Statistical Methods for Geography: A Student's Guide,* 2nd edn, London: Sage.

Verzani, J. (2004) *Using R for Introductory Statistics,* Boca Raton, FL: Chapman and Hall/CRC Press.

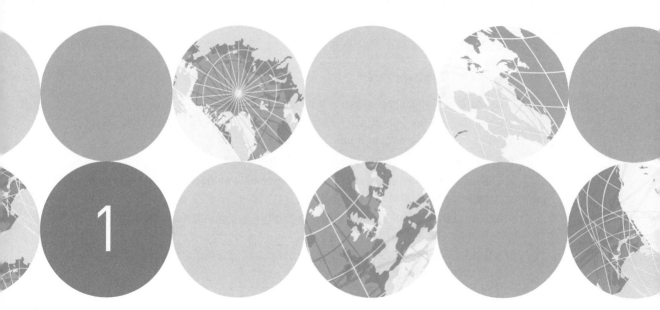

Data, statistics and geography

Chapter overview

The aim of this chapter is to convince you that studying statistics is a good idea for any student of geography, environmental science or geographical information science.

Our argument is that data collection and analysis are central to the functioning of contemporary society. It follows that knowledge of data handling and of statistics is a necessary skill to contribute to social and scientific debate.

We present statistics as a reflective practice, a way of approaching research that requires a clear and manageable research question to be formulated, a means to answer that question, knowledge of the assumptions of each test used, an understanding of the consequences of violating those assumptions, and awareness of the researcher's own prejudices when doing the research.

At their simplest, statistics are used to form basic numeric summaries of the processes, events or activities that the data represent. Yet, this is only the starting point for analysis. Statistics can also go beyond a sample of data, to help decide whether information gleaned from the sample is true more generally. Statistics ask if one 'thing' is related to another, or if the one causes the other.

We do not claim that statistics are the only way of doing research. But, they are important tools for validating theory, making predictions, helping to make sense of the world, engaging in policy research, offering informed commentary about social and environmental issues, and to help make the case for change.

Learning objectives

By the end of this chapter you will be able to:

- Appreciate the importance of studying statistics as a student of geography, environmental or geographical information science.
- Define what is meant by geographical data and analysis.
- Give a working definition of the difference between geographical and non-geographical forms of analysis.
- Summarise some debates surrounding the use of numbers and statistics in human and social geography.
- Have an appreciation of the impact of spatial association when using statistics for geographical enquiry.

1.1 Statistics: a brief introduction

Students tend to be uneasy about statistics. This is hardly surprising. Statistics involve the language of mathematics, of formulae and notation – a language that is mystifying to new learners and worryingly intolerant of mistakes. Concepts such as 'degrees of freedom' are less intuitive than they ought to be and frequent use of the Greek alphabet is like shorthand for 'keep away!' Yes, learning statistics is a challenge. But the problem is deeper than that: statistics seem innately off-putting:

> [E]very year it seems that the majority of students, who apparently grasp many non-statistical concepts with commensurate ease, struggle to understand statistics.
>
> (Dancey and Reidy 2004, p.1)

Of course, there are exceptions. It is a generalisation to say, 'Students tend to be uneasy about statistics.' We have not met all students to ask them. Even if we had, each individual's response to the question 'do you like statistics?' would depend on a range of factors including time of day, their understanding of the last stats class, peer pressure, other work commitments, whether they thought a 'yes' would get us to go away, and so on.

However, we are not suggesting *every* student is uneasy about statistics. We know there is variation between each individual's attitude and aptitude. Our observation is simply of a *general trend* and is based on experience. Admittedly, there is a risk that our experience is unusual. There could be widespread fondness for statistics in places other than the Universities of Bristol and Leicester (the institutions in which we work). If so, our generalisation is wrong. But that would surprise us. We believe we have encountered enough sufficiently typical students to remain confident in our assertion that 'students tend to be uneasy about statistics'.

Is that how you feel? If so, you are not alone. That said, if you understand all you have read so far, then you might be closer to understanding statistics than you previously imagined. Because, in the preceding paragraphs, we laid out a basic principle of statistics: to learn from a *sample* (of students we have met), something we have confidence will be generally true for other members of the *population* too (the student population, though we have not met them all), whilst at the same time being mindful of variation (the fact that some students like statistics more than others). The formulae of statistics may seem abstract and confusing, yet all they really do is define and quantify what is meant by confidence and variation, offer ways of detecting a trend, establish how general the trend is, and offer tools to describe and explain what we have found out.

Ultimately geographical data analysis and statistics are about ideas and concepts. These matter far more than the formulae do. The concepts are ways to think about and do research. It is far better you 'get behind' these ideas than get stuck in the quagmire of equations. It is easy enough to download free statistical software that can do the calculations for you.

However, we are not convinced that a complete avoidance of equations really helps you learn. Whilst a computer may do the calculation, to know when and why it is made remain important. When you are familiar with their language, equations and notation are the most succinct and least ambiguous way of expressing calculations and concepts. As authors, our responsibility is to help you learn. Nevertheless, you can breathe easy for a while – you will not find any equations in this chapter! Instead, we discuss what we mean by geographical data analysis, make a case for why studying statistics is important (even if you have no intention of using them ever again) and consider some of the practical problems involved in employing statistical techniques in geographical research.

1.2 Why you should study statistics

A good question is 'why?' Why do you need to study statistics? Are they *really* relevant to your learning, to your research plans?

Reasons for human geographers

Questions about the use of numbers and statistics in, especially, human geography have been influenced by the 'cultural turn' of the social sciences and humanities. We address some of the issues later in this chapter. For now we make a simple appeal to the more philosophically sceptical student. Data collection and analysis are central to the functioning of contemporary societies, underpinning systems of science, governance and production. This is evident by listening to a news channel: have the Dow Jones indices risen or fallen; what is the crime rate in your locality; how many pupils passed examinations this year; what impact did that have on school 'performance tables'; has the government met targets on reducing carbon emissions; who is topping the baseball league; what are the waiting times at your local hospital; was this the wettest or hottest summer on record; how long should we wait before saying that an outbreak of seasonal flu has passed . . . ? It follows that knowledge of statistics is an entry into debate, informed critique and the possibility of change.

Consider the following text that appeared on the website of the new economics foundation (nef), 'an independent think-and-do tank that inspires and demonstrates real economic well-being':

What gets counted, counts. nef is redefining approaches to value and measurement so that those things that matter most to people, communities and to achieving a sustainable planet are made visible and measurable.

Practices of measurement and valuation are still often focused narrowly and on the short term. Sometimes things that are easy to count, outputs, are the things that get measured and thereby valued. Instead nef believes measures should be focused on outcomes and how lives, communities or the environment changes as a result of policy.

(www.neweconomics.org)

This is a very positive agenda and one we embrace. At the same time, we recognise that some see counting as part of the problem: it classifies, it separates, it labels in ways that can reflect narrow political interests. However, to withdraw to a position where lack of knowledge prevents informed criticism is utterly self-defeating: it leaves the status quo free from effective challenge.

As Barnes and Hannah (2001, p.379) note:

geographers should take numbers and statistics seriously [. . .] because they are a crucial component in the construction of social reality.

We need not like the reality and can raise important questions about how, why and for whom our worlds are 'organised, controlled, manipulated, studied, and known' (Surveillance Studies Network 2006). But it is best to do so from an informed position. We cannot enter the debate without the requisite knowledge and understanding to do so. In short, there is little sense in standing on the sidelines, even if learning statistics is not an enticing prospect. Dorling (2003, pp.369–370) understands this, writing:

Statistics are duller than ditch water. In and of themselves they tend to be of interest to people who are not very interesting. [!] To me, it is only when statistics are set in wider context that they begin to come to life [. . .] Suppose you are interested in the issue of poverty. Poverty rates are statistics [. . .] If you are interested in why higher levels of poverty persist in some places and not others [. . .] you are unlikely to get far without the numbers and methods which make up statistics.

Presented clearly, statistical information can catch the media's eye. Here is what the BBC reported about Thomas and Dorling's (2007) social atlas of British society, on Saturday 8 September 2007 when the book featured as a leading news item:

CLASS SEGREGATION 'ON THE RISE'
A UK social atlas suggests that British society is becoming more segregated by class, researchers have said [. . .] It found that:

- An average child in the wealthiest 10% of neighbourhoods can expect to inherit at least 40 times as much wealth as a typical child in the poorest 10%
- In some areas, 16-to-24-year-olds are 50 times more likely to attend an elite university than in others
- In the most impoverished parts of the country young adults in this age group are almost 20 times more likely not to be in education, employment or training than those in the wealthiest neighbourhoods

- There are no large neighbourhoods where under-five-year-olds from the highest social class spend time with any other class of children other the one just beneath them.
(http://news.bbc.co.uk/1/hi/uk/6984707.stm, after Thomas and Dorling, 2007)

People tend to be cynical about statistics but they can be used to support social action. An obvious example is how they are enacted in the debate about climate change. Statistics predicting temperature increases between 1.1 and 6.4°C over the next century, and research indicating that 2°C is the threshold beyond which there are dangerous impacts to nature, humans and the global economy, are dramatic reminders of the potential impact of global warming (Intergovernmental Panel on Climate Change 2007). They have caught the public eye and been used to mobilise political support for the reduction of carbon emissions. But, like many statistics, they have also proved contentious and some groups have rallied against them.

Reasons for GI scientists

For the user of geographical information systems, we seek to complement the textbook introductions to geographical information science. Those textbooks tend to focus on technologies to collect geographical data, show how that information is encoded and visualised by computer systems, and explain how useful knowledge can be obtained by querying, linking and manipulating the data in particular ways.

The amount of data 'out there' is enormous and increasing rapidly (but with uneven geographical coverage). Neogeography has emerged as a term partly about how geospatial and Internet-based technologies such as Google Maps can be used to share and to display geographical data, without expert training in GIS or cartography (see, for example, Turner 2006).

Whatever the merits of this 'new geography' it benefits from a shot of the old. As a baseline, the added extra of studying statistics is to address the uncertainties and ambiguities of using data analytically. A more progressive reason for studying statistics is because of the increasing integration of mapping capabilities, the visualisation of data and of statistical analysis in free-to-download software packages (Bivand *et al.* 2008; Rey and Anselin 2006).

Reasons for all readers

The benefit of studying statistics is in gaining a skill set that is transferable to other research methods, disciplines and walks of life. In many countries there is a recognised 'shortage of well developed quantitative skills across the social sciences and a pressing need to build capacity' (www.esrc.ac.uk). Careers seeking statistical expertise include actuarial work, marketing consultancy, environmental assessment, policy research, careers in government, and many more.

Studying statistics encourages an approach to research that is reflective, thoughtful and mindful of the limitations of data and their analysis. Encouraging the researcher to form a clear and manageable research question, a means to answer the question, and awareness of the assumptions, methodological limitations and the researcher's own prejudices is a discipline conducive to all empirical work, quantitative or qualitative.

For statistical research, it is good practice to ask whether the results have both statistical *and* substantive meaning. Focusing on the statistical helps to avoid sensationalising events that are either potentially random or entirely predictable – to avoid claiming they are unusual when they are not.

Focusing on the substantive gets away from mechanical thinking and blithely assuming that because an analytical result passes some statistical test it must therefore be meaningful. Ziliak and McCloskey (2004, p.544) put it nicely when concluding their review of papers published in the *American Economic Review*: 'we should be economic scientists, not machines of walking dead recording 5% levels of significance'. (We see why they refer to 5% later in the book.)

Consider, for example, the prevalence of geographical data handling technologies making it easy to 'mine', query or simplify large databases of complex information to discover interesting patterns and associations. Because there is so much data we expect to discover *something*. Chance is like that: given enough opportunity the statistical test will be passed. But that does not mean the result has any social, scientific or real-world meaning necessarily. One problem is that there is 'too much' data:

> In any scientific investigation, there are bound to be some sources of bias. Perhaps the sample wasn't quite random, or maybe the respondents tended to overestimate their income. We do everything we can to minimize those biases, but we can never eliminate them entirely. Consequently, small, artifactual relationships are likely to creep into the data. A large sample [or dataset] is like a very sensitive measuring instrument. It's so sensitive that it detects those artifactual relationships along with the true relationships.
>
> (Allison 1999, pp.58–59)

Another is the risk of circular thinking. If too great an emphasis is placed on data collection and processing, and too little on a guiding socio-economic or scientific theory, it risks the 'building up of ad hoc generalizations' (Anyon 1982, p.34), of simply 'fitting the facts' to meet prior but unwarranted expectations.

Key concept 1.1 Don't believe everything you read!

It is good to question what a statistic is telling you. Something can appear statistically important yet still have no social or scientific meaning. Be guided by theory and by prior research findings, as well as by what the data are saying. Try not to 'fit the facts' to the data, just because it is convenient to do so. Statistics are an aid for your intelligence and judgement, not a replacement for them!

1.3 Types of statistic

To this point we have considered statistics in a general sense. In thinking more deeply about their application in research, it is helpful to identify three uses for statistics.

The first is **descriptive,** providing a summary of a set of measurements. Averages may be calculated, the most common values identified and the spread of values obtained (for example, by finding the minimum and maximum).

They can be used for making simple comparisons. For example, the (mean) average annual rainfall at Darwin airport in Northern Australia was 1847.1 mm for the period 1971–2007. For Melbourne, on the south coast, it is lower: 654.4 mm. (Source: Australian Government Bureau of Meteorology www.bom.gov.au).

Descriptive statistics are useful in their own right and as a precursor to more sophisticated analysis. They are cognitively useful: they help to frame understanding of the data and of the real-world events and processes they measure. They are also useful diagnostically, especially for detecting error (often caused by missing data due to incomplete record keeping). If the average rainfall for Melbourne were −10 mm we would need to check the data carefully as it clearly cannot be negative! We look more at descriptive statistics in Chapter 2.

The second use is *inferential*. This was discussed at the beginning of the chapter. Inferential statistics go beyond describing a set of data in its own right and use the information to say something about the population from which the data were sampled.

The UK Conservative Party (as part of the governing coalition) is predicted to have the support of 37% of the electorate against 41% for the Labour Party. How does a polling company know this when it asked only 1500 people from an electorate of approximately 45 million? It does not know, not for certain, but it can make an inference if it believes the sample is representative of the electorate as a whole. Being representative means those who have been polled are not disproportionately located in, say, Labour-voting industrial areas or Conservative-voting rural areas. The poll is not biased towards a particular section of the electorate.

The polling company is not claiming that levels of party support will be exactly the same in the population as in the sample. Some difference is inevitable. What the company can reasonably discount, however, is the possibility of the Conservative and Labour Parties having equal support amongst the electorate. The 4 percentage point difference found in the sample is too great for that to be likely. In studies of this type, a percentage point difference of about 3 percentage points or above suggests the parties are unlikely to have equal support from the electorate. We will look at inferential statistics again, in Chapters 5 and 6 particularly.

The third use is **relational** and potentially **explanatory**. At a minimum they identify whether particular events or circumstances coincide. Does shaving less than once a day really increase a man's risk of having a stroke (BBC News online, 7 February 2003)? Unlikely, but they could well be related, perhaps by lifestyle. Going further, relational statistics become explanatory when they offer firm insight on what causes what. In an analysis of rural to urban migration in China, Wang and Fan (2006) identify some of the personal, social and economic reasons that can lead a migrant not to continue in a city but instead return home to the countryside. Chief amongst them is age (the older the migrant, the more likely they are to return) but important also is family responsibility – those who are married and/or have children are increasingly likely to return.

Going further, if statistics are explanatory then they may also be predictive. A statistical model might be built to answer 'what if?' questions. Relational and explanatory statistics are the focus of Chapters 7 and 9.

> **Key concept 1.2 Types of statistic**
>
> There are three types of statistic: those that **describe** and summarise a set of data in its own right; those that 'go beyond' the data to **infer** something more general about the population from which the data were sampled; and those which **examine relationships**.

1.4 Analysis and error

Having given an overview of statistics, we now turn to a related word: analysis. The *Oxford English Dictionary* gives one definition of it: 'a detailed examination of the elements or structure of something'.

For statistical analysis we examine data – information that can be stored, retrieved, queried, summarised, classified and used for calculations by a statistical package running on a computer. The data will often be numeric but not always. We could, for example, use a word processor to count how many times the word statistics appears in this book. The result is numeric but the objects of the analysis are words.

When we examine data we do so because we are interested in what the data represent. If they are measurements, then our interest is in what has been measured. If I measure the heights of my children, I am not merely interested in the numbers. I am interested in what those numbers tell me about my children's heights (for example, how much they have changed since the last time I made a measurement).

The problem is that any measurement is subject to **error**: my children not standing straight, the tape measure warping, my not aligning it vertically, and so forth. An analogy is often used from physics and engineering. Consider a radio wave carrying an analogue transmission of a news broadcast. A number of factors affect the quality of what you hear, including the quality of the radio, the size of its aerial, the distance from a transmitter, atmospheric conditions, and so forth. What you want to hear is the news – the signal. Any background hiss is unwanted noise. A challenge of statistical analysis is to separate 'the noise from the signal'.

Imagine we used a questionnaire-based survey to gather data about business and manufacturing for a study about regional economics and competitiveness. What we have to accept is that the data obtained are imperfect. People may misunderstand the questions or make an error when completing a tick box. We may miscode the data or ask questions that are not fully relevant. Nevertheless, we hope to take the data and use them to understand why some regions prosper whilst others decline.

There will always be noise in data. Take a ruler and measure the length of the following: _____. How long is it? Can you repeatedly get the same *exact* answer twice? Probably not, because it depends on exactly where you position the ruler, where your eye adjudges to be the end of the line, the ruler being perfectly aligned with the line, a lack of hand shake, no creasing of the paper, and so forth.

Obviously it has length. And you can probably measure it well, just not definitively because you cannot avoid the chance of error. Also, you need to be wary of systematically repeated error: for example, measuring the line in centimetres but thinking they are inches; or using a plastic ruler that has warped in the Sun.

Because noise – error – creeps into any analysis, no single answer, measurement or result will be perfectly definitive (and you would have no way of knowing even if it was). Put another way, if an experiment is repeated twice or more, there is no guarantee the same *exact* result will be obtained each time. Instead, we expect the opposite: to find variability in the results even if the structure and properties of that being analysed do not change.

However, error is not the only reason to expect data to vary. Social, economic and environmental systems themselves create outcomes that are not the same everywhere. That is geography! Mean temperature in New York is not the same as in Melbourne. Property prices in Manhattan are not the same as in The Bronx. Atmospheric pollution in Los Angeles is different to that in Hawaii. Acknowledging this variation does not mean the issue of error goes away. It adds to the challenge of explanatory data analysis: to find and explain real variations, not those due to error.

Key concept 1.3 Error is unavoidable

Error creeps into any data collection and analysis – it cannot be avoided because there are so many ways it can arise: a faulty instrument; misreading; data wrongly coded; 'rounding' errors (for example, treating 1.7 as 1); tiredness; interference (for example, poor atmospheric conditions); misunderstanding the research question; and so forth. The hope is that the effects of error will be neutral overall – that they will not seriously distort what is learned from the data. The worst errors are those that are hard to detect, difficult to correct and have a large impact on the analysis.

1.5 Geographical data and analysis

Much of this book uses real-world data to demonstrate statistical ideas and methods. And many of those data are geographical.

Geographical data are records of what has happened at some location on or near the Earth's surface. The key characteristic of a geographical data set is that it records both pieces of information: what has happened (sometimes called the attribute value) and where (the location or georeference).

For many of the methods presented in this book, location is just a way of categorising (grouping) data. For example, we may be interested in regional differences in health across a country, using standard statistical methods to examine whether the age-adjusted rate of heart disease in the southern states of the United States (6.7%) is significantly different from the western states (6.0%). The answer is not especially (see Lethbridge-Çejku *et al.* 2006, Table 2, which reports on the US National Health Interview Survey, 2004).

Although we may describe this as geographical data analysis we should not confuse it with the geographical analysis of data. The distinction is in the methods of analysis used. The test comparing southern and western states could also be used to compare heart disease for males and females (they are higher for men: 8.3% vs 4.9%) or to examine the effects of poverty (the rate of heart disease for those with family incomes less than $20 000 is 9.2%; for those earning $75 000 or more it is 5.3%). Whilst it can be used to test for geographical differences, there is nothing exclusively geographical about the test itself.

In fact, most of the statistical tests in this book actually assume there are no geographical patterns in the data. If that seems a peculiar assumption for a geographer to make, you would be right. However, it does make the statistics easier to calculate in much the same way that it is easy to calculate there is a one-in-two chance of a tossed coin landing head up if the result of that toss is unbiased and in no way dependent on any preceding tosses, and rather harder to calculate if the chances of landing a head are dependent on previous tosses in complex and not easily observable ways.

However, for some of the analyses, both attributes and location matter. These generate genuinely geographical statistics. They not only work with geographical data, but are explicitly designed for them. They recognise that what makes geographical data special is not just their attributes but also their locations. Therefore those locations are not discarded. These tests belong to the fields of geospatial and spatial statistics, which are introduced in Chapters 8 and 9. The spirit of both is encapsulated by **Tobler's 'first law of geography'**: 'everything is related to everything else, but near things are more related than distant things' (Tobler 1970, p. 236). Some tests seek to validate the law (which is really an assumption). Others incorporate it into the method of analysis.

Key concept 1.4 Geographical data analysis

Geographical data are those that record a measurement and also the location for the measurement. **Geographical data analysis** can be understood as using such data to detect, examine, understand or predict events and phenomena that are of relevance to a geographical study.

However, the methods of analysis may not themselves be geographical. Many of the more common descriptive, inferential and relational statistics make no specific allowance for where the data were collected. It is only later in the book that we encounter genuinely geographical methods of data analysis.

1.6 Some problems when analysing geographical data

Whist advocating a statistical way of working, we want to be honest about its limitations. In this section we consider three issues: spatial autocorrelation; the problem of determining causes; and some more general critiques there have been of 'quantitative geography'.

Spatial autocorrelation

Geographers look for, study and explain patterns of association found on the Earth's surface. They ask why things come together *where* they do. The suggestion is that 'things' found near to each other would display similar characteristics or behaviours. They are associated with each other.

A reason they might is because they are exposed to the same formative processes and conditions for change. These include the base conditions that lead things to form in particular places. They also include interaction effects, the reciprocal relationships between one thing and others, and their dependency on each other to grow or to evolve.

An example of this is plant succession and the ways that particular types of plant species form communities in particular locations. The vegetation that grows is a function of the ground conditions and other environmental determinants that are external to the community but which act upon it. It is also a function of the community itself, of how the growth of one plant helps or hinders another.

A similar idea, found in the sociological and economic literature, is of 'neighbourhood effects' operating within the places where humans live. It is reasonable to suppose that in politically liberal societies, economic systems, including the effects of the housing and job markets, and other more behavioural factors such as cultural ties, will place constraints on where people choose to live. These are the 'ground conditions'. Once there, people interact. Hence, one understanding of a neighbourhood effect is what happens when a particular group of people come together in a neighbourhood, influence each other's behaviour and so begin to 'stamp' a mark on that community in a way that could become self-perpetuating in terms of other people the neighbourhood attracts.

A simple example appears outside the window of the writer's house. The front lawns of nearby properties are being converted, one after another, into off-road parking for cars, and all by the same contractor. This change is a function of the physical properties of the neighbourhood (narrow roads), of the suburban, family-oriented lifestyles of the people who live here, and of people seeing the conversions and thinking 'oh, we would like that, too!'

A more profound example is a downward spiral in terms of crime, social cohesion, welfare dependency, and so forth affecting a community. However, neighbourhood effects are difficult to measure and standard (regression) techniques are not ideally suited for doing so. (An alterative method – multilevel modelling – is considered in Chapter 9.)

If what we measure in one place is related to what we measure nearby then we have what is called **spatial autocorrelation** – a pattern in the measurements (in the data values). What this amounts to saying is 'geography matters' and is the sort of statement that a geographer likes to hear! Unfortunately, the more traditional methods of analysis are founded on the reply 'it shouldn't!' because of the assumptions of independence they make about the data. This takes us back to the distinction between non-geographical and geographical forms of analysis.

In general, the presence of *positive* spatial autocorrelation appears more prevalent in geographical studies and is the basis for Tobler's 'first law of geography' (see Section 1.5 and also Key concept 1.5). It means finding similar values for data collected at locations that are close together. However, it is also possible to have *negative* spatial autocorrelation,

> in which the neighbors of locations with large values tend to have small values, the neighbors of locations with intermediate values tend to have intermediate values, and the neighbors of locations with small values tend to have large values.
>
> (Griffifth 2006, p.336)

For example, negative spatial autocorrelation will arise if children living in mixed neighbourhoods separate along ethno-cultural lines in terms of the schools they attend; that is, if students choose schools that are disproportionately characterised by students of their same ethno-cultural background (Harris and Johnston 2008).

Nevertheless, although we may caution about the use of non-geographical forms of analysis in geographical research, in many cases those methods remain useful. The practical consideration is whether the assumptions of a statistical test have been sufficiently met for the interpretation of the results to be correct. To this end, it will be useful to map the data, look for any geography and, if found, keep it in mind in the analysis and in the interpretation of the results.

Key concept 1.5 Spatial autocorrelation and the 'first law of geography'

The **first law of geography** is attributed to Waldo Tobler and states that everything is related to everything else, but near things are more related than distant things. This is a statement about **spatial autocorrelation**.

Positive spatial autocorrelation means that if you take measurements of some geographical feature or phenomenon, similar values will be found for data from locations that are situated close together. Finding that the values of neighbours oppose each other is evidence of **negative spatial autocorrelation**. In either case, there is a spatial (geographical) dependency within the data that violates the assumption of independence in many statistical tests.

Determining what causes what

A second difficulty when using statistics to model, explain, predict or describe social and natural systems is that the systems are complex, dynamic, interacting and therefore changing. Rarely is there the equivalent of a fume cupboard in which we can simply add sodium to water and produce a flame (please don't try this, it can also be explosive). Experimental conditions are hard to obtain: it is difficult if not impossible to take social and environmental systems and, whilst holding everything else steady, add only one thing, the effects of which we want to understand. Data often are collected after the event so are used to look at what has already happened, not to test directly what would happen if . . . ; and social data frequently are recycled – they are collected for reasons other than the specific purposes of any particular piece of research.

Because of all this, whilst it is relatively straightforward to find that one thing is *associated* with another, to prove what *causes* what is a far more challenging undertaking because there is a lot that can confound (confuse) the analysis.

For now we note the effects of scale. A core problem is that what we see in the world is dependent upon the scale we use to view it. This presents something of a 'catch-22' – we can use analytical methods to draw out geographical patterns and differences but what we draw out is dependent upon the geographies we used to look at those patterns in the first place.

It is a little like using Google Earth: what we see depends on the viewing scale, except most data sets do not offer the option to zoom in or out but are fixed at predetermined scales such as zip codes, census tracts, regional authorities, grid squares or land parcels that are not necessarily optimal for the analysis. They therefore place a limit upon it and affect what we see in the results. Two particular problems are the modifiable areal unit problem (MAUP) and ecological fallacy. These are discussed in Chapter 8.

Some critiques of quantitative data analysis

We have already touched on debates in human geography and other areas of social science about using numbers for research and study. They are important so we consider them in more detail here. If you are new to such debates, you might wish to skip this section for now and return later on.

The debates follow a period during the 1960s and early 1970s when quantitative methods were particularly fashionable. They express a concern that a discipline solely shaped by statistics and model building risks blunting its critical edge (by being naive, complacent or complicit about why and for what purposes data are collected), will miss those aspects of society and/or humanity that are not simply captured by social surveys and data sets (for example, human emotion, spirituality, feelings, identity), and, by focusing on the general, lacks sensitivity to difference.

We share these concerns, though we note that the idea that quantitative geography cannot be 'critical' is misguided (Barnes 2009). It is also a shame that discussions about them are too often strident, a situation well summed up by Barnes and Hannah's (2001, p.382) tongue-in-cheek comment that 'quantitative methods tend to be automatically dismissed by their dissenting critics as positivist. And positivism, as everyone knows, is evil and wrong.'

Here it is sufficient to understand positivism as encapsulating the traditions of scientific discovery: principles of experimentation and logic that use founding axioms ('rules') and previously obtained knowledge to form new ideas that are, in turn, proven right or wrong based on that which can be observed or recorded (therefore verified or falsified). In this way, theories are refined and knowledge steadily accumulates within organising frameworks and unifying laws. For example, we can (at the risk of oversimplification) establish a lineage from Aristotle through to Galileo, Newton and Einstein whereby understandings of the laws of motion and of time are said to progress towards a comprehensive understanding of the fabric of our Universe.

Clearly there is an association between science, numbers, measurement and statistics. However, these should not be connected in simplistic or singular ways. Especially, we need to avoid conflating statistics and quantitative methods with narrow interpretations of the practices of science, a mistake that some textbook introductions to geography continue to make.

A problem with simple characterisation of how science operates is that it risks portraying science as a unified whole, within which there is harmony on how to 'do science' and on what constitutes fact and theory. Yet, we know there is controversy and disagreement. Consider, for example, debates about theoretical physics and ideas such as string theory that has mathematical coherence but is unproven in a classic, experimental sense.

Our point is not the sociology of science argument that science (necessarily) takes place within human contexts and networks, that these must influence how, when, why and for whom it operates, and therefore science cannot be said to be perfectly objective if this means to be free from all human emotion, whim or bias. Although that argument is valid and important, what we are warning against is simple caricatures of how and why people do research. For example, it is not unusual to see data handling technologies such as geographical information systems (GIS) described as positivistic:

> given the demand for GIS and quantitative geography in the public and private sector, it is likely that unreconstructed, positivistic geography is secure for the foreseeable future.
>
> (Kitchin 2006, p.28)

However, this coupling of GIS, quantification and positivism is problematic, not least because uses of GIS are frequently data driven and interpretive – exploring data to discover knowledge, not answering or testing pre-formulated hypotheses.

This may seem pedantic, but recognising diversity prevents confusing methods with the motivations of researchers in the way the following authors have:

> Spatial scientists tend to reduce methodology to technique, being bothered about the correct running of a test but less about anything entailed in the deriving of the data [. . .] nor anything following conceptually, politically, ethically or otherwise from choosing to tackle the data statistically rather than some other way.
>
> (Cloke *et al.* 2004, p.21)

We agree whole-heartedly with the need to consider the conceptual, the political and the ethical. However, the quote above gives a misleading impression of how and why quantitative methods are employed in geographical research. Without wishing to be combative or territorial, it is important to recognise that 'spatial scientists' have had a long-standing and continuing interest in such issues.

For example, the Radical Statistics Group (yes, really!) was formed as long ago as 1975 for researchers and statisticians who share a common concern about the political implications of their work and an awareness of the actual and potential misuse of statistics (www.radstats.org.uk). In 1998, the National Center for Geographic Information and Analysis formed initiative 19, looking at the social implications of how people, space and the environment are represented in GIS (a body of literature now usually referred to as Critical GIS). In that same year, Esnard (1998) documented some 'abuses and questionable uses' of GIS in local governance and offered some 'Portable,

Provisional Codes' to serve as a simple set of ethical guidelines. The GIS Certification Institute (GISCI) requires that its Code of Ethics be signed as a condition of certification for a GIS professional (see www.gisci.org/code_of_ethics.aspx). Similarly, the Royal Statistical Society has a Code of Conduct which it expects of its members. In recent years, 'qualitative GIS' have emerged, influenced by understandings of how geographical data and analysis are gendered in particular ways, and seeking to incorporate less structured, less official and less formal modes of representation in geographical data handling technologies (Harris 2007; Kwan and Knigge 2006).

In short, there are '-isms' other than positivism that are consistent with a statistical way of working (for example, realism, critical realism, pragmatism and feminism, see Aitken and Valentine 2006). Researchers need not think that they are revealing natural, definitive and enduring laws of society, to believe that their methods can help understand some of the structures and processes that shape and are shaped by societies, and by individuals and groups within them. They need not believe in their complete objectivity, impartiality or lack of bias, to still believe that the knowledge they are forming is valid, and useful. They need not believe that their data and methods are beyond debate, to believe they contribute to debate. And they need not privilege quantitative methods as *the* way of working, to recognise their value as *a* way.

Key points

- Statistics have three main purposes: to be descriptive, inferential or relational.
- Knowledge of statistics is important because data collection and analysis are central to the functioning of society.
- Statistical practice encourages a research rigour and reflectivity that is useful for both quantitative and qualitative research.
- Users of statistics do so for a variety of reasons and with a range of motivations, beliefs and philosophical perspectives.
- Geographical data analysis is about detecting, examining, understanding and predicting events and phenomena that are geographical in nature. It tries to separate what is true of those features from error in the data.
- It is a good idea to ask whether the results of research have both statistical and substantive meaning, probing beneath the surface of what the data appear to be telling you and making links to academic debate and theories.
- Problems when using geographical data include the modifiable area unit problem, the ecological fallacy and the difficult in determining causation within complex and changing systems.
- Tobler's first law of geography states that everything is related to everything else, but near things are more related than distant things. This is a statement about spatial autocorrelation.

References

Aitken, S. and Valentine, G. (2006) *Approaches to Human Geography,* London: Sage.

Allison, P.D. (1999) *Multiple Regression: A Primer,* Thousand Oaks, CA: Pine Forge Press.

Anyon, J. (1982) Adequate social science, curriculum investigations, and theory. *Theory into Practice,* 21(1), 34–37.

Barnes, T. and Hannah, M. (2001) The place of numbers: histories, geographies and theories of quantification. *Environment and Planning D: Society and Space,* 19(4), 379–383.

Barnes, T.J (2009) 'Not only . . . but also': quantitative and critical geography. *Professional Geography,* 61(3), 292–300.

Bivand, R.S., Pebesma, E.J. and Gomez-Rubio, V. (2008) *Applied Spatial Data Analysis with R,* New York: Springer.

Cloke, P.J., Cook, I., Crang, P. and Goodwin, M.A. (2004) *Practising Human Geography,* London: Sage.

Dancey, C. and Reidy, J. (2004) *Statistics Without Maths for Psychology,* 3rd edn, Harlow: Prentice Hall.

Dorling, D. (2003) Using statistics to describe and explore data. In N. Clifford and G. Valentine (eds) *Key Methods in Geography,* London: Sage, pp.369–382.

Esnard, A. (1998) Cities, GIS, and ethics. *Journal of Urban Technology,* 5(3), 33–45.

Griffifth, D.A. (2006) Hidden negative spatial autocorrelation. *Journal of Geographical Systems,* 8(4), 335–355.

Harris, R. (2007) Qualitative analysis. In *Encyclopedia of Geographical Information Science,* London: Sage, pp.355–357.

Harris, R. and Johnston, R. (2008) Primary schools, markets and choice: studying polarization and the core catchment areas of schools. *Applied Spatial Analysis and Policy,* 1(1), 59–84.

Intergovernmental Panel on Climate Change (2007) *Climate Change 2007 – The Physical Science Basis: Working Group I Contribution to the Fourth Assessment Report of the IPCC (Climate Change 2007): Working Report of the IPCC,* 1st edn, Cambridge: Cambridge University Press.

Kitchin, R. (2006) Positivistic geographies and spatial science, In S. Aitken and G. Valentine (eds), *Approaches to Human Geography,* London: Sage, pp.20–28.

Kwan, M. and Knigge, L. (eds) (2006) Theme issue: qualitative research and GIS. *Environment and Planning A,* 38(11), 1999–2074.

Lethbridge-Çejku, M., Rose, D. and Vickerie, J. (2006) Summary health statistics for US adults: National Health Interview Survey, 2004. *Vital Health Statistics,* 10(228). Available at: http://www.cdc.gov/nchs/data/series/sr_10/sr10_228.pdf, accessed 20 October 2010.

Rey, S.J. and Anselin, L. (2006) Recent advances in software for spatial analysis in the social sciences. *Geographical Analysis,* 38, 1–4.

Surveillance Studies Network (2006) *A Report on the Surveillance Society,* London: Information Commissioner's Office. Available at: http://www.ico.gov.uk/upload/documents/library/data_protection/practical_application/surveillance_society_full_report_2006.pdf, accessed 20 October 2010.

Thomas, B. and Dorling, D. (2007) *Identity in Britain: A Cradle-to-grave Atlas,* Bristol: Policy Press.

Tobler, W.R. (1970) A computer model simulation of urban growth in the Detroit region. *Economic Geography,* 46(Supplement: Proceedings. International Geographical Union. Commission on Quantitative Methods), 234–240.

Turner, A. (2006) *Introduction to Neogeography,* Sebastopol, CA: O'Reilly.

Wang, W.W. and Fan, C.C. (2006) Success or failure: selectivity and reasons of return migration in Sichuan and Anhui, China. *Environment and Planning A,* 38(5), 939–958.

Ziliak, S.T. and McCloskey, D.N. (2004) Size matters: the standard error of regressions in the American Economic Review. *Journal of Socio-Economics,* 33(5), 527–546.

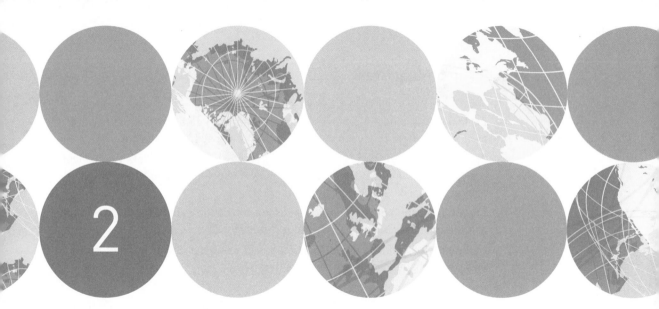

Descriptive statistics

Chapter overview

This chapter is about 'everyday statistics', the sort that summarise data and describe them in simple ways. They include the number of home runs this season, average male earnings, numbers unemployed, outside temperature, average cost of a barrel of oil, regional variations in crime rates, pollution statistics, measures of the economy and other facts and figures that are all around us on websites, in newspapers, on TV, and so forth.

These are the sorts of descriptive information that come about by observing and measuring something, then by summarising the data in clear and straightforward ways. We look at simple ways to summarise the properties of a data set numerically in a table or graphically in a chart. We focus on two of the most important summaries – measures of what is commonly called the average and measures of how much the individual data values vary around that average.

Appearing in this chapter are examples of formulae and notation. Arguably they are unnecessary because statistical packages will do the calculation and produce an answer at the push of a button. The reason for their inclusion is not because they need to be memorised or even ever used to calculate the statistics by hand. The reason is that they express a concept and that concept *does* need to be understood.

Please don't be put off by the equations. Our advice is that you focus on learning the concepts and then see how they are formalised by the equations. Learn the ideas, not the formulae.

Learning objectives

By the end of this chapter you will be able to:

- Explain why it is useful to have an understanding of data types and be aware of how they affect the types of analysis that are possible for the data.
- Appreciate that 'the type' is not an immutable property of the data themselves but relates to what is being measured, how it is being measured and why it is being measured.
- Show how sorting data into numeric order allows the minimum, the maximum and the middle value of the data to be determined.
- Outline the advantages and disadvantages of using bar charts or histograms as a way of visualising the shape of a data set.
- Produce a summary table of a data set.
- Explain why 'the average' is a vague term and give exact definitions of three measures of central tendency: the mean, the median and the mode.
- Describe measures of spread around the centre including the range, the interquartile range and the standard deviation.
- Identify the median, the interquartile range and unusual data values using a box plot.

2.1 Data and variables

Forming knowledge from data is central to statistical analysis. Data are information obtained by measuring 'things' at different times or locations across a study region. Each measurement is called an observation, because it is a record of what has been observed where and when the measurement was made. That 'thing' – the focus of the study – will be a social, scientific or environmental feature, or process, that we want to learn more about. Geographical examples include elevation, surface temperature, ozone levels, soil quality, unemployment, noise, traffic congestion, access to shops and amenities, voting behaviour, water salinity, slope stability, vegetation cover, life expectancy, environmental quality, quality of life, happiness, crime rates, . . . amongst many, many others.

Here, when we talk about data, we are referring to **numeric data**. Archives, manuscripts, photographs and paintings are all sources of data for particular types of research but they are not numeric. Furthermore, not all data that appear numeric originate that way. The word statistics is clearly a combination of letters yet it can still be coded as numbers (for example, 10, 19, 20, 1, 20, 9, 19, 20, 9, 3, 19, indicating there are 10 letters and giving their position in the alphabet). In this book, data are shorthand for numeric data, and numeric data are those that arise from a process of measurement.

A single observation cannot measure change or variation, and there is the risk of the measurement being wrong. Therefore a series of observations is made by recording measurements for different times or places within the study region. The set of measurements is known as a **variable**. It is called a variable for good reason. Looking at the observations will reveal variations in the measurements – they are not all the same.

There are two reasons for the variations. First, because what is measured is not stationary – it has different characteristics at different times and places. For example, unemployment is greater in some places and in some years more than others; the same is true of air pollution. Secondly, there are other things that degrade the quality of the measurement, creating error, that do themselves vary across time and space.

Given both these considerations it would be surprising if all observations had *exactly* the same value. The starting point for statistics is to offer simple descriptions of each variable. Later in the book we are more ambitious, wanting to understand why things vary and going beyond description to look for causes.

Key concept 2.1 Data and variables

Data are measurements of something of interest. They are also called observations because the measurements help us to observe (and to quantify) an attribute of whatever is being studied.

A set of measurements is called a **variable** because the values are unlikely to have the same value at all times and locations at which the measurements were made. Descriptive statistics summarise key information about the variable. Other types of statistics seek to explain the causes of the variations.

Data types

It is important to distinguish between different types of numeric data because what you can do with data is related to the types of data they are. This is true for more basic descriptions of data and for more advanced statistical methods. However, thinking about data types is not only about matching the data to an appropriate method of analysis – though that is important. To consider the data type is also to give thought to the properties of the phenomenon under study and to the sorts of data it can generate, as well as to how the phenomenon has been conceptualised and observed by the researcher. The data type is a product of what is measured, how it is measured and why it is measured.

One of the most common ways of defining data types is by using the Stevens scale (Stevens 1946). This is discussed in many introductory textbooks, is useful and appears later in this chapter. However, it may also be too strict to apply to real-world data and, in any case,

> scale type, as defined by Stevens, is not an attribute of the data, but rather depends upon the questions we intend to ask of the data and upon any additional information we may have. It may change due to transformation of the data, it may change with the addition of new information that helps us to interpret the data differently, or it may change simply because of the questions we choose to ask.
>
> (Velleman and Wilkinson 1993, p.69)

Instead, we begin as Dancey and Reidy (2004) do by distinguishing between discrete and continuous data. Discrete data take on one from a limited (and therefore finite) set of possible values. Because of this it is possible to count how many times each specific value appears in the data and to produce a tally – a frequency table – of those counts.

Discrete values tend to be whole numbers, also known as integer numbers. These are values which have no fractional part and are written without a decimal point. For example, 1 is an integer whereas 1.1 is not. The value −2 is also an integer but −2.0 is not: the inclusion of the decimal point implies −2.0 is from a set of non-integer values that might include −2.1 or −2.2, and −2.0 might itself be an approximation of −2.04 for example.

In contrast, **continuous** data are drawn from an infinite set and can take on any value – or, at least, any value between a lower and upper limit. They are often 'real' or 'floating-point' numbers, those with a decimal point. For example, 1.001 is a real number, written to three decimal places (there are three digits after the decimal point). Another is −4.112 34, with five decimal places.

Because there are an infinite number of values continuous data could take, it is futile to produce a frequency table for them. Many or all of the values that do appear will do so only once. Instead, we can arrange them into groups (for example, 0–4.99, 5–9.99, 10–14.99, etc.) and then count the number of members in each group. The result will not be independent of the groupings. Later in the chapter we will see that the way we group continuous data affects our portrayals of them.

It could be argued that no data are perfectly continuous since there are always limits to the precision (the number of digits) by which events can be measured and recorded. No measurements are drawn from a truly infinite set. However, the difference between discrete and continuous data is better understood as a property of what is being measured than of the data themselves. For example, most light switches have two discrete states, either on or off, whilst the luminance of energy-saving light bulbs increases, on a continuous scale, from when they receive an electric current to when they are fully lit.

It is also better to understand the words discrete and continuous as two ends of a continuum along which different sets of data are placed. It then becomes a matter of asking 'which is the better model for the data I have, what I want to do with it and for what I am trying to study?' Integer data are never really continuous, but unless the set contains only a few (discrete) values then it is neither unusual nor necessarily mistaken to treat it as if it were continuous.

An exception to our focus on numeric data will be **categorical variables** – those that are labels distinguishing one category from others. For example, M and F, for males and females; 1, 2 and 3 for different types of land use (1 = arable; 2 = forest; etc.); 'left' and 'right' for political allegiance; low, moderate and high for levels of risk; 'good' and 'bad' as qualitative judgements; and so forth. We will use categorical variables to split a set of measurements into groups that are then compared. For an introduction to explicitly categorical data analysis see Agresti (2007), or Wrigley (2002) for a geographical perspective.

> **Key concept 2.2   Discrete and continuous data**
>
> As this chapter progresses, have in mind the difference between discrete and continuous data.
>
> **Discrete** data are those that take on one from a restricted set of possible values. They are usually whole or integer data.
>
> **Continuous** data could take on any value, or any value within a lower and upper limit and to a certain level of precision. They are 'real' or 'floating-point' numbers.
>
> In practice, the difference between these data types is not fixed or immutable. Discrete data are often analysed as though they are continuous, and continuous data can be grouped into discrete categories (for example, by sorting height data into 'short', 'average' and 'tall').

2.2   Simple ways to make sense of and present data

It is time to stop talking about data and to get our hands dirty with some! Table 2.1 gives the reported number of human rabies cases in the United States for each of the years from 1974 to 2005. Rabies is a virus that affects the central nervous system and is fatal if not treated. For brevity, we will call these 'the disease data'. The data have been chosen for their simplicity. The few years of data, the fact that rabies is a rare disease and the fact that there can only be a whole number of cases make them relatively straightforward to commence with. They are discrete data.

The data are in chronological order. Reading from left to right, then from top to bottom, Table 2.1 gives the number of reported cases for each of 32 separate years. We can say that $n = 32$, where n is shorthand for the number of observations. It is sometimes called the length of the variable.

The data are few but it still takes a while to look at Table 2.1 and to understand what it is showing. Table 2.2 is easier to comprehend. It has the number of cases sorted in ascending order. It is easy to see that there were no reports of human rabies in 6 of the 32 years, and that the highest number was recorded in 2004, when there were 7 cases.

Table 2.1   The number of reported cases of human rabies in the United States for each of the years 1974–2005

Year	1974	1975	1976	1977	1978	1979	1980	1981
Cases	0	2	2	1	4	4	0	2
Year	1982	1983	1984	1985	1986	1987	1988	1989
Cases	0	2	3	1	0	1	0	1
Year	1990	1991	1992	1993	1994	1995	1996	1997
Cases	1	3	1	3	6	5	3	2
Year	1998	1999	2000	2001	2002	2003	2004	2005
Cases	1	0	4	1	3	2	7	2

Source: Centers for Disease Control and Prevention (2007).

Table 2.2 The number of reported cases of human rabies in the United States, 1974–2005, sorted in ascending order. The lowest, highest and middle values are highlighted

Year	1974	1980	1982	1986	1988	1999	1977	1985
Cases	0	0	0	0	0	0	1	1
Year	1987	1989	1990	1992	1998	2001	1975	1976
Cases	1	1	1	1	1	1	2	2
Year	1981	1983	1997	2003	2005	1984	1991	1993
Cases	2	2	2	2	2	3	3	3
Year	1996	2002	1978	1979	2000	1995	1994	2004
Cases	3	3	4	4	4	5	6	7

Source: Centers for Disease Control and Prevention (2007).

This gives us the minimum value (min. = 0), the maximum (max. = 7) and the difference between them, which is called **the range** (here also equal to 7).

We can use Table 2.2 to find a middle value lying halfway between the minimum and maximum number of cases. It is the cell (the value in the table) that has as many observations to one side of it as it does to the other and is called **the median**. There is a problem, though. Because our n is an even number ($n = 32$), no such cell exists. We can select the 16th value, which has 15 values less than it and 16 values greater, or the 17th value, which has 16 less and 15 greater. Neither is exactly midway. As it happens, both values are the same and we obtain a median of 2 in either case. Were they different then one option would be to pick one, perhaps at random (making the decision by the equivalent of a coin toss). Another would be to split the difference, calculating 'the average' of the two (specifically, the mean average; see Section 2.5).

Having produced Table 2.2, it requires little effort to count how often any particular value appears within it. That information can then be used to produce a frequency table and to establish a typical number of rabies cases per year. For example, the value of zero appears six times in Table 2.2, one appears eight times, two appears seven times, and so forth.

Table 2.3 is the frequency table. It shows there was most commonly one case of rabies per year in the United States during the period 1974–2005. This, the most frequently occurring value, is called **the mode** or modal average.

In the first column of the table, the letter x means simply 'a value', and letter f is an abbreviation of frequency – the number of times each value appears in the source data. We can say that for $x = 0$, $f = 6$; for $x = 1$, $f = 8$; for $x = 2$, $f = 7$; and so on.

Tables are useful but graphics are often preferred for their visual impact and because they can be better at providing an overview of the data, though often at the expense of

Table 2.3 Frequency table for the reported number of human rabies cases within the United States during the years 1975–2005. The most frequent number (the mode) is 1, occurring for 8 of the 32 years

x	0	1	2	3	4	5	6	7
f	6	8	7	5	3	1	1	1

Source: Table 2.2, using data from the Centers for Disease Control and Prevention (2007).

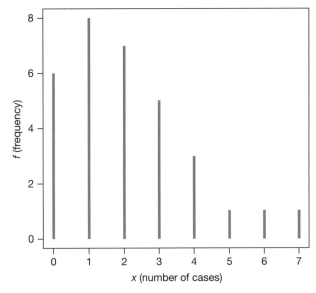

Figure 2.1 Bar chart showing the distribution of the disease data

specific detail. Graphics can also be more persuasive in their ability to 'sell' a particular message, which also leads to the possibility of deliberate distortion. Good practice is to use graphics to summarise key features of the data clearly and accurately – to support the analysis but not to conceal or exaggerate certain aspects of it. For a classic treatise of how (not) to lie with statistics and graphics, see Huff (1954).

Figure 2.1 is the graphical equivalent of the frequency table. It is a **bar chart** in which the height of each bar is proportional to the number of times the value shown on the horizontal axis appears in the disease data. From it we can again determine that the modal average is 1. Figure 2.1 gives a more obvious impression of 'the shape' of the disease data than does Table 2.3. It shows the data spread out between the minimum of zero and the maximum of seven but are skewed towards to the lower values, meaning they occur more frequently than higher ones.

Key concept 2.3 The minimum, the maximum, the median and the range

If a set of data is arranged in numeric order from the lowest to the highest values then it is easy to find the **range** and also the **median**. The **minimum** is the lowest value, the **maximum** is the highest. The range is the difference between the two. The median is the middle value, the one that has as many other values to one side of it as it does to the other.

The summary table

Having explored the data a little, it is helpful to present what we have learned in a clear and concise way. In books, journal articles and scientific reports it is rare for a data set to be published in full. The reason is simple: to do so is both cumbersome and fairly

Table 2.4 Summary of the disease data: reported number of human rabies cases within the United States during the years 1975–2005

n	Minimum	Maximum	Mean	Median	Mode
32	0	7	2.09	2	1

Source: Centers for Disease Control and Prevention (2007).

pointless; there usually are too much data to interpret in their 'raw' form. Instead, the characteristics of the data are summarised, often in a summary table.

Table 2.4 reports the number of observations (n), the minimum and maximum values, the median and the mode for the disease data. It therefore contains information about the middle and most frequently occurring values, and about the spread – the range – of the data. It is quite rare to see the mode reported in a summary table (because most data are continuous and the mode is not appropriate) but here it says something useful: it reports the most typical number of human rabies cases each year.

Also included is the mean, an average number discussed further in Section 2.5. We have written it to two decimal places (2.09), which seems sensible for the data. We could write it to more, 2.093 750 0 for example, but that gives a level of precision, an exactness that is difficult to justify. It would imply a great deal of certainty in the data: that every case has been identified, that there is no migration of infected persons across national borders, that the result is not affected by the reporting period (such that an infection in late December, albeit somewhat unlikely, might actually be diagnosed the next year), and so forth.

Finally, the source of the data should be and is fully acknowledged beneath Table 2.4.

2.3 Some useful notation

We mentioned in Chapter 1 that one of the barriers to learning statistics is the notation and the formulae that come with them. We said that the ideas and concepts employed in data analysis are more important than how they are expressed using strange signs and symbols and we meant it! However, we also suggested, 'when you are familiar with their language, equations and notation are the most succinct and least ambiguous way of expressing calculations and concepts'. That also remains true.

We already have introduced some notation into this chapter. The letter n has been used as shorthand for the number of observations, the length of the variable. We have used f as an abbreviation of frequency and the letter x has indicated 'a [any] value'. However, x could also be understood as 'a variable', in which case subscripts can be used to reference specific observations within a set of data. The notation x_1 refers to the first observation. Similarly x_2 refers to the second observation, and so forth.

Table 2.5 makes this clearer. It shows the raw disease data, the same as in Table 2.1, except it refers to the first year – the first observation – as x_1, the second year (the second observation) as x_2, the third as x_3, and so on to the nth value, which is x_{32}. It says that $x_1 = 0$, $x_2 = 2$, $x_3 = 2$, etc. The point is not that Table 2.5 is better than Table 2.1 as a way of presenting the data (if anything the original is better because it retains

Table 2.5 Demonstrating the use of subscripts to refer to specific elements (observations) within a set of data. See text for explanation

	x_1	x_2	x_3	x_4	x_5	x_6	x_7	x_8
Cases	0	2	2	1	4	4	0	2
	x_9	x_{10}	x_{11}	x_{12}	x_{13}	x_{14}	x_{15}	x_{16}
Cases	0	2	3	1	0	1	0	1
	x_{17}	x_{18}	x_{19}	x_{20}	x_{21}	x_{22}	x_{23}	x_{24}
Cases	1	3	1	3	6	5	3	2
	x_{25}	x_{26}	x_{27}	x_{28}	x_{29}	x_{30}	x_{31}	x_{32}
Cases	1	0	4	1	3	2	7	2

information about the years). The point is the use of x and a subscript to refer to a specific part of the data. Note that the last value, x_{32}, can also be referred to as x_n because $n = 32$ for the disease data.

As well as using subscripts to point to individual values in the data, it is usual to embellish the x in ways that refer to some overall summary of the data. For example, $\tilde{x}$ is sometimes used to refer to the median value, whilst the mean is often indicated by $\bar{x}$.

You should be aware that any notation is somewhat arbitrary. Some is broadly fixed by intuition or convention but you will still encounter variation between authors. Our choices are fairly orthodox but you should still anticipate differences between this and other textbooks. It is another good reason to learn what the symbols represent rather than exactly how they are written.

2.4 A second data set

We will now use some of the techniques and notation we have learned to summarise a second data set. However, we will also introduce some new approaches because this set is more complicated than the first: it has more observations and a greater variety of values. The data are more continuous than before but are still limited to whole numbers.

Table 2.6 summarises the data which are about the air quality for 715 counties in the United States. The data give the number of days per county, in 2007, when ozone was the main air pollutant (as opposed to carbon monoxide, ozone, sulphur dioxide or particulate matter). They are from the Environmental Protection Agency's AirData reports (http://www.epa.gov/air/data/). The values range from a minimum of 0 days to a maximum of 362 days with ozone commonly the main pollutant for about half the year: the median value for the 715 counties is 184 days, the mean is 190 days and the mode is 215 days.

Table 2.6 Summary of the air quality data: number of days in 2007 when ozone was the main air pollutant, as reported for each of 715 US counties

n	Minimum	Maximum	Median, $\tilde{x}$	Mean, $\bar{x}$	Mode
715	0	362	184	190	215

Source: the Environmental Agency 2008: http://www.epa.gov/air/data/.

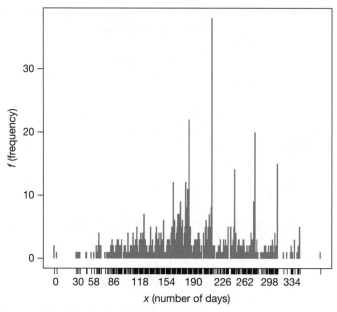

Figure 2.2 Bar chart showing the distribution of the air quality data: number of days when ozone is the main air pollutant in 715 US counties

Figure 2.2 is a bar chart for the air quality data. It provides useful information about the distribution of the air quality data but a problem is emerging. Now there is a much greater variety of values we are getting towards the point where there are too many distinct values to count each separately on the chart or in a frequency table.

The obvious solution is to group the data so that similar values are counted together. One way to do this is to organise the data into categories. One possible scheme is shown in Table 2.7 and in Figure 2.3.

Table 2.7 Creating categories for how often ozone was the main air pollutant in each of 715 US counties

Category	Number of days	Count
Very infrequently	0–49	9
Infrequently	50–99	56
Below average	100–149	127
Average	150–199	218
Above average	200–249	162
Often	250–299	103
Very often	300–349	39
Nearly always	> 350	1

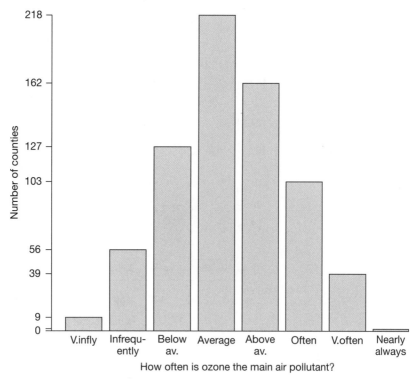

Figure 2.3 Bar chart showing the number of counties in each of eight categories based on the air quality data

Histograms and rug plots

The categories ('very infrequently', 'infrequently', and so forth) do not add anything to the air quality data especially, other than giving them a qualitative label that may aid understanding. There is no particular need for the categories. Figure 2.4 plots the data without the categorisations but with the groupings.

Whereas Figure 2.3 was a bar chart, Figure 2.4 is a **histogram**. They look very similar. The convention is to use bar charts for categorical data (or data that have been categorised) and histograms for more continuous data.

In a histogram, the groups are called 'bins' and are like containers into which the data are put. In Figure 2.4, the groups have an equal bin width, meaning the distance along the horizontal axis of the histogram is always the same from one group to the next. In Figure 2.4, each bin begins 50 units from the last. Having an equal width is usual for a histogram.

The histogram gives an impression of how the data spread out, of the values that are common in the data and those that are not. It suggests how the data are distributed. Unfortunately that impression is not neutral but is affected by the number, the width and the positioning of the bins.

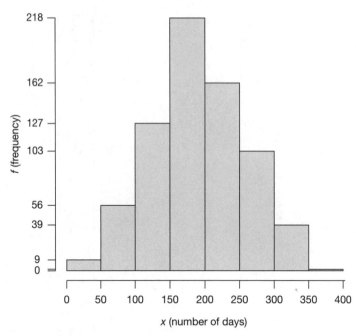

Figure 2.4 Histogram showing the distribution of the air quality data when grouped into eight bins of equal width

Figure 2.5 shows the distribution of the air quality data when the number of bins is increased to 19 or decreased to 4. The data have not changed but their shape now looks different. This example warns against being *too* reliant on purely visual summaries of data but it is not meant to imply that histograms should never be used. They, like other forms of data visualisation, are extremely helpful aids for understanding data and detecting error. It just requires care about how they are interpreted.

A second problem with a histogram is that the individual data values are lost within each bin – they are subsumed into the group. Admittedly, this is an ungrateful criticism: the reason for using the histogram was the need to group the data. However, it is a criticism worth making since it is possible to have the best of both worlds. Figure 2.6 includes a **rug plot** where 'threads' are drawn out from the bins and on to the horizontal axis giving an indication of where the specific values lie (Tufte 2001). Note, for example, that the bin of values ranging from 0 to 49 (days) actually contains no observations with a value from 5 to 29.

Stem and leaf plot

Another way of showing a distribution in a way that can retain the individual data values is to use a stem and leaf plot. Figure 2.7 shows a plot of the ozone data. The stem is the vertical axis which, in this example, increases in units of 10 but is marked (more sparingly but perhaps confusingly) as 0, 1, 2, etc. This means the 1 is actually 10, 2 is

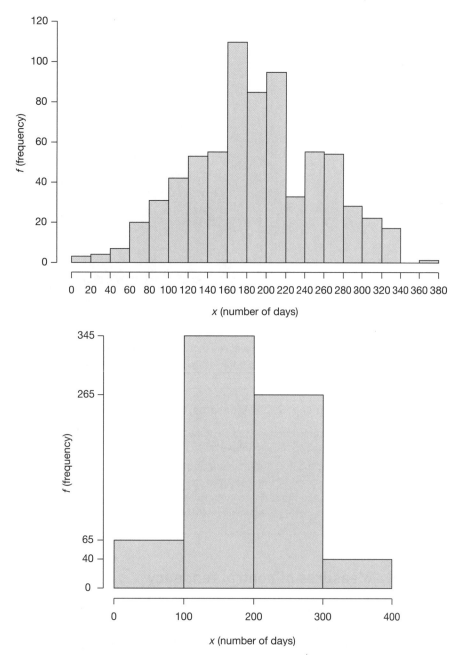

Figure 2.5 Bar charts showing the distribution of the air quality data when grouped into 19 bins of equal width (upper graph) and 4 bins of equal width (lower graph)

actually 20, and so forth. The leaves are the numbers going from left to right across the page, here increasing in units of 1. Reading from top to bottom and from left to right, the stem and leaf plot is giving the values in the ozone data sorted into numeric order: 0, 0, 0, 4, 30, 32, 34, 35, 44, 45, 51, 55, 58, 58, 59, etc.

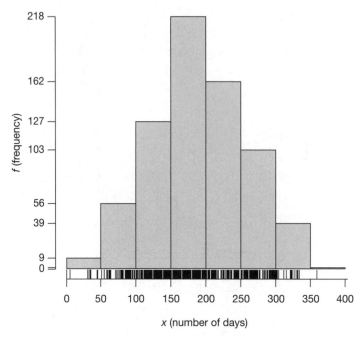

Figure 2.6 A histogram and rug plot showing the distribution of the air quality data: number of days when ozone is the main air pollutant in 715 US counties

---

Key concept 2.4 Histograms, rug plots, and stem and leaf plots

Histograms can be used to show the shape or the **distribution** of a set of continuous (or quasi-continuous) data. They reveal the range of the data and suggest the values that are most common occurring within it.

The shape of the histogram depends not only on the data but also on the way they are grouped together into bins. Within the bins, the specific data values are hidden. To address this problem, consider using a **rug plot** or a **stem and leaf plot**.

2.5 Measures of central tendency

Looking back at Figure 2.6, it can be seen that the ozone data have a humped distribution with a 'clumping' of values near the centre of the horizontal axis. It is more common for counties to have 150–199 days with ozone as the main air pollutant than to have 100–149 or 200–249 days. Values below 100 or above 249 days are even less frequent. The further the values are from the centre of the graph, the less common they become.

But where, actually, is the centre? Looking at Figure 2.6 you might reasonably say it is between 150 and 199 days, whereas the upper plot of Figure 2.5 might lead us to suggest a value of between 160 and 179 days. In either case, the answer is vague. The way we typically define the centre of the data is by calculating an average, usually the mean average.

```
 0 | 004  ◄──── This line says there are two zeros in the ozone data and one 4
 1 |       ◄
 2 |       ◄──── These lines say there are no data values in the tens or twenties
 3 | 0245 ◄──── This line records the following values in the ozone data: 30,32,34,35
 4 | 45   ◄──── A value of 44 and a value of 45
 5 | 15889
 6 | 1122223344
 7 | 024688999
 8 | 0112455666788899
 9 | 001223335667788
10 | 0111124557788889  ◄────  There is one value of 100 in the data, four values
11 | 00112223444455677789999      of 101, one of 102, one of 104, two of 105, etc.
12 | 000122223333333444555667789
13 | 00112222233445556788999
14 | 0000011222333444455666677788999
15 | 00000012344555666677788899
16 | 000011111223333333333334444455555566677777888889999999
17 | 00000111111122223333333334444455555566778888899999
18 | 0000000000001111111122222222222333334444444444444444444445566666788
19 | 001122233444555666677778889
20 | 011333345566666677777788899
21 | 0000011111112222233344444444555555555555555555555555555555555555555577
22 | 1244556668
23 | 011122333455677777
24 | 000001223344455555566666666666677778
25 | 0011122244556677788899
26 | 012234556678888999
27 | 0111111111233333333333333333333456667
28 | 13466889
29 | 0112344566777889
30 | 00001122344444444444444444
31 | 26
32 | 12234777
33 | 012244444
34 |
35 |
36 | 2  ◄──── A single value of 362
```

Figure 2.7   An annotated stem and leaf plot showing the distribution of the air quality data: number of days when ozone is the main air pollutant in 715 US counties. The stem is in units of 10, the leaves are in units of 1

The mean

The mean is what most people call 'the average'. That average is found by taking a set of data, adding the values together and then dividing by the number of values there are. More strictly, the mean is the arithmetic mean, being one from a family of means (which include the geometric and harmonic means, see Crawley 2005). However, 'the arithmetic' is often dropped from its title. When we refer to the mean we are talking about the **arithmetic mean**.

The definition of the mean is simple to translate into a formula. We have already introduced x to indicate a value and n to say how many values there are. All that is needed is an extra symbol, Σ, which means 'add together'. The symbol is the upper case of sigma, our first use of the Greek alphabet! Continuing to use $\bar{x}$ to denote the mean, then

$$\bar{x} = \frac{\sum x}{n} \qquad (2.1)$$

This simply says 'the mean ($\bar{x}$) is obtained by adding all the values together ($\sum x$) and then dividing the total by how many numbers there were (n)'. So, if there were five numbers, 1, 4, 8, 5 and 2, then the mean average is four:

$$\bar{x} = \frac{\sum x}{n}$$
$$= \frac{1+4+8+5+2}{5} = \frac{20}{5} = 4 \qquad (2.2)$$

Another way of writing Equation 2.1 is as follows:

$$\bar{x} = \frac{\sum_{i=1}^{n} x_i}{n} \qquad (2.3)$$

The difference is subtle but draws on the use of subscripts introduced in Section 2.2. Equation 2.3 is just a little more specific, saying sum the values from the first observation (x_1, which is what x_i becomes if $i = 1$) to the nth (x_n, when $i = $ n). The use of the letter i is arbitrary; we could equally well use j, k or any other letter.

The mean is not the only measure of **central tendency** – we have already introduced two others. One was the mode, which can be useful in identifying the most frequently occurring value in a set of discrete data. The other is the median, the middle value. Of the three, the mean is the most widely used. However, it is susceptible to unusually low or high values ('outliers') that can have an undue influence on it, moving it away from what would otherwise be the centre.

If we imagine a number line with lower numbers to the left and higher numbers to the right, the mean will be pulled to the right by the presence of an unusually high value within a set of data. It will be pulled to the left by a lower value. The mean can be likened to a pivot point, balancing the numbers to one side against those to the other. If the numbers 'move' in one direction then the pivot must move too, to keep the balance. A simple illustration is given by Figure 2.8. Note that in the example the middle value does not move. This suggests why the median is more robust (less sensitive) to the influence of unusually high or low values than the mean is and would be a better measure of central tendency when such values occur.

If the mean is susceptible to the effects of unusually high or low values, why is it so widely used? There are two reasons. First, it is easy to calculate. Now, that may surprise you: to find the middle value (or the one most frequently occurring) seems simpler than 'adding all the values together and then dividing the total by how many numbers there are'. However, that is to confuse a concept with its calculation. It takes longer to sort a long list of values from lowest to highest than it does to add them together and divide by n. The point may seem mundane but much of statistics was developed before the advent of computers that do the sorting for us.

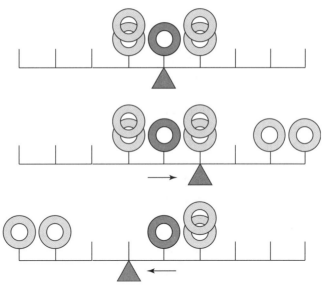

Figure 2.8   The mean is like a pivot point balancing the values on one side with the values on the other. It is moved by the presence of outliers in a way that the median is not

Secondly, the mean has useful mathematical properties. For a discussion of these see Bolstad (2007, pp.40–41).

Trimmed mean

A variant of the mean that compensates for unusually high or low values is the trimmed mean. The idea is simple. First, the data are sorted into numeric order. Next, a chosen proportion of the data is removed (trimmed) from each end of the sorted list. These will include the highest and lowest values. The mean is then calculated for the remaining data.

The value of the trimmed mean depends on the proportion of the data that are removed. For the ozone data, the trimmed mean is 191 days if 5% of the data is cut from each end of the sorted list. This is little different from the actual mean of 190 days and suggests the data do not contain any unusually high or low values.

Key concept 2.5 Measures of central tendency

Measures of central tendency, more often referred to as averages, include the **median**, the **mode** and the (arithmetic) **mean**. The median is at the middle of a sorted list of values, the mode is the most frequently occurring value and the mean is the result obtained by 'adding all the values together and then dividing the total by how many numbers there are'. Because of this, the mean is susceptible to unusually high or low values that could pull it away from the centre of the data. In such cases, a **trimmed mean** could be used instead.

2.6 Some measures of spread and variation

A measure of central tendency is useful but says nothing about how the data are distributed around it. Knowing only an average is not, by itself, especially helpful.

Field (2003) gives an example of why not. Imagine measurements are taken of a pollutant in each of three streams on 10 different days. The first stream is found to have a mean pollution concentration of 6.1 units, the second stream has 9.1 units and the third has 5.6 units. If the threshold beyond which a stream becomes toxic is 10 units, then which is the stream to worry about?

Intuitively the answer is the second stream because its mean (of 9.1) is closest to 10. But that intuition could be misleading without knowledge of how the individual measurements vary around it. In Field's example, the second stream is, in fact, never toxic. It is close to being so but still never exceeds the threshold. In contrast, though stream C has the lowest mean, it also exhibits the greatest spread of values, becoming toxic on two occasions. The example is summarised by Figure 2.9. It is stream C, not stream B, that poses the greatest risk.

In general, a measure of central tendency should always be accompanied by a measure of **spread**. It should be possible to identify whether all the values are packed close to the centre or whether they are more variable. The minimum and maximum values provide a clue for this by defining the range. However, it is a rather crude measure of spread if the minimum and maximum represent extreme values that are infrequently occurring and

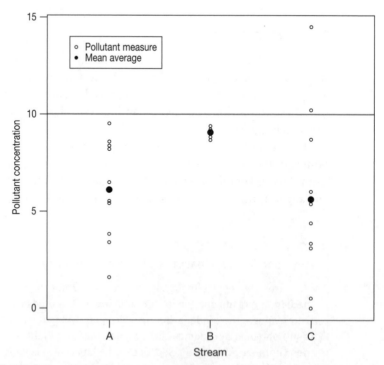

Figure 2.9 Why it is unwise to focus on an average alone. Although stream C has the lowest mean, it is also the only stream that breaches the pollution limit of 10 units

Source: The example is from Field (2003).

therefore atypical. It would be better to focus on the more commonly occurring values and define a range using only those instead. That is what the interquartile range does.

Key concept 2.6 Averages can be misleading

Measures of central tendency should be accompanied by measures of the spread around them. If there is little spread then an average such as the mean will give a realistic impression of the specific values from which it was calculated. However, if the spread is great then any specific value could be quite different from what we expect given the mean. Any misimpression could result in poor decision making such as underestimating a risk.

The interquartile range

The interquartile range (IQR) is calculated in a way similar to finding the median. For the median, the data are sorted into numeric order. The median lies at the midpoint of the sorted data and can be used to split the set into two, each containing half the data. For the IQR the data are sorted and quartered.

The first quarter of the data is called the first quartile and the value marking *the end* of it is sometimes abbreviated as Q1 (one-quarter along). The last quarter of the data is called the fourth quartile and the value marking *the beginning* of it can be abbreviated as Q3 (three-quarters along).

The difference between Q3 and Q1 is the interquartile range, literally meaning between quarters. In effect, the IQR trims off the first and last quarters of the data, leaving two-quarters – one-half – remaining. This is the middle half of the data and will exclude the highest and lowest values. The IQR is a measure of the 'mid-spread' of the data.

Quantiles

The IQR is based on the division of an ordered data set into four quartiles, each containing one-quarter of the observations. However, there is no particular reason why data should be split into four. In principle they could be divided by any number, provided it is less than or equal to the number of observations, *n*.

Quintiles arise when the data are divided into five groups and deciles when they are split into ten. Percentiles are when groups contains one-hundredth of the data. Quantiles is the generic name for these and for other possible divisions.

Key concept 2.7 Quantiles

Quantiles are found by sorting data into numeric order and then splitting the set into a number of equal-sized groups (or approximately so, if they cannot be split exactly). Splitting the data into four groups produces **quartiles**, each group containing one-quarter of the data. Splitting the data into five produces **quintiles**, each containing one-fifth of the data, 10 groups gives **deciles**, each containing one-tenth, and 100 groups gives **percentiles**, each containing one-hundredth.

The standard deviation

As we have seen, the range considers the how far the minimum value is from the maximum, whilst the IQR looks at the difference between the first and third quarters of the data. An alternative way of summarising the spread is to calculate how far *each* individual observation is from the centre of the data and then calculate an average.

This is what the standard deviation does. It is abbreviated by the letter s and is defined by the formula

$$s = \sqrt{\frac{\sum_{i=1}^{n}\left(x_i - \bar{x}\right)^2}{n-1}}$$

(2.4)

To understand the logic behind the formula, it will now be derived slowly and in stages. It is worth taking time to read these stages carefully because the standard deviation is widely used in statistical work and employs concepts that we will encounter again in later chapters, specifically the '**sum of squares**' and '**degrees of freedom**'.

Stage 1: Calculating the sum of squares

Consider again the set of numbers 1, 4, 8, 5 and 2, the mean of which is 4. The first observation has a value that is three below the mean. Using notation, $x_1 - \bar{x} = 1 - 4 = -3$. The second observation has a value that is equal to the mean: $x_2 - \bar{x} = 0$. The third has a value that is four above the mean, and so forth.

One way to summarise the differences could be to add them together:

$$\sum_{i=1}^{n}\left(x_i - \bar{x}\right)$$

(2.5)

Unfortunately the idea is flawed because the mean is the pivot point whereby values less than the mean balance out those that are greater. Consequently, Equation 2.5 will always give a value of zero. Put another way, the negative values cancel out the positives, and this suggests the solution: ensure there are no negative values! They are removed by multiplying each difference by itself, $(x_i - \bar{x}) \times (x_i - \bar{x})$, which is $(x_i - \bar{x})^2$. In other words, the differences are squared.

We are now interested in the sum of squared differences or, more simply, the 'sum of squares', which is

$$\sum_{i=1}^{n}\left(x_i - \bar{x}\right)^2$$

(2.6)

Key concept 2.8 The sum of squares

The amount of spread in a set of data can be calculated by finding the difference between each individual value and the mean, squaring each difference, and then summing the squared differences together. The result is called the **sum of squares**.

Stage 2: Creating an average (finding the variance)

The sum of squares describes the total variation of observations around their mean. If the observations are packed tightly around the mean then the sum of squares will be lower than if those observations are more spread out. However, it is not only the spread of the data that determines the sum of squares; it is also dependent on the number of observations. As n increases, there are more observations to vary around the mean; therefore, the sum of squares must increase. We cannot say one set of data is more variable than another simply because one contains many more observations than the other.

We need a more meaningful way to compare data sets that differ in length. This is achieved by taking an average: in principle, by dividing the sum of squares by the number of observations, n. In practice, the more usual division is by $n-1$. Whilst it rarely makes much difference, the reason for preferring the latter relates to the 'degrees of freedom' (see Key concept 2.9) and also to the distinction between a sample and the population from which the sample is drawn (see Chapters 4 and 5). Either way, the result is a measure of the average squared variation around the mean. It is called **the variance** and is given the symbol s^2:

$$s^2 = \frac{\sum_{i=1}^{n}\left(x_i - \bar{x}\right)^2}{n-1} \tag{2.7}$$

Key concept 2.9 Degrees of freedom

The **degrees of freedom** are the amount of flexibility you have to change the values of some observations if certain properties of the data set are fixed. Those properties are often referred to as parameters.

Imagine somebody takes some playing cards and, having removed the picture cards, looks at 10 cards at the top of the pack. That person then tells you the sum of the 10 cards is sixty before dealing them face down. The question is: what is the maximum number of cards you need to turn up before you know the face value of them all?

The answer is nine. You know there are 10 cards and you know the sum of their values. By deducting the values of the first 9 cards from sixty, you can determine the value of the final card.

Alternatively, consider a data set containing $n=10$ observation for which the mean is known and must stay constant. You can take any nine of the observations and change their values to anything you like. But, having done so, you have no choice about what the tenth element equals: its value is fixed by the mean and by the other observations. Your 'degrees of freedom' are limited to $n-1$ of the data values.

Be aware that the number of degrees of freedom is not always $n-1$. It depends on the parameters of the data set required for a particular test or measure. A formal definition of degrees of freedom offered by Crawley (2005, p.37) is **the sample size, n, minus the number of parameters, p, estimated from the data**.

Stage 3: Returning to the original units of measurement

The variance is widely used to summarise the spread of observations around their mean. A disadvantage is that it is not measured in the same units as the data.

Consider again the air quality data. The variance in the number of days when ozone is the main source of air pollution is 4373 units, around the mean of 190 days.

The variance is measured in units that are the square of the original measurement units and which arise from the sum of squares. It is therefore on a different scale to the original data and is why the variance appears much greater than the mean. It would be more intuitive if it and the data had the same measurement scale.

To achieve this, the 'square root' of the variance is calculated. The square root is indicated by the symbol $\sqrt{}$ and the square root of the variance is **the standard deviation**, s,

$$s = \sqrt{\frac{\sum_{i=1}^{n}\left(x_i - \bar{x}\right)^2}{n-1}} \qquad (2.8)$$

This takes us back to where we started. Note that the standard deviation and the variance are directly related – they are the square and the square root of each other. This is clear from the notation where the standard deviation is indicated by s and the variance by s^2. Consequently, their interpretation is the same: both are measures of the average spread around the centre of a data set and increase as the spread does. The standard deviation of the ozone data is 66 days, which is the square root of the variance of 4373.

Key concept 2.10 The standard deviation and the variance

Both the **standard deviation** and the **variance** are summary measures of the amount of variation found in a set of data.

The variance, s^2, is the sum of squares divided by the degrees of freedom. The standard deviation, s, is the square root of the variance. The standard deviation has the advantage of having the same measurement units as the data and can be loosely interpreted as a measure of average spread.

2.7 Presenting the centre and spread of data

Box plots

Box plots are an effective way of displaying the median and the IQR of one or more variables. They also show the minimum and maximum values, and draw attention to especially high or low values. Box plots can be drawn either horizontally or vertically. Horizontal plots can be easier to interpret because they are read from left to right and then downwards, the same way you are reading this book.

Figure 2.10 is a box plot of the air quality data, with a rug plot drawn below it. The thick, vertical line near the centre of the rectangular box marks the median value. If you were to extend that line downwards until it crossed the horizontal axis, it would do so at a value of 184 days (you can confirm this is the median by looking back at the summary table, Table 2.6). If you extend lines downwards from the left and right borders of the box they will be found to cross the horizontal axis at the first and third quartiles, 147 days (Q1) and 240 days (Q3), respectively.

Extending outwards from the rectangular box are '**whiskers**'. Conventionally these extend on the left side to the observation that is closest in value but not less than the first quartile minus 1.5 times the IQR (Q1 − 1.5 × IQR), and on the right side to the observation that is closest to but not exceeding the third quartile plus 1.5 times the interquartile range (Q3 + 1.5 × IQR). As such, they cannot extend past the minimum or maximum values in the data but may stop at them. Where they do not reach as far as the minimum and maximum values, the observations that lie beyond the whiskers are possible outliers, so named because they are spread out away from the other observations. The whiskers provide a 'rule of thumb' for identifying low or high values and to consider whether they are unusual. If they are then it may not be appropriate to analyse them with the rest of the data. It could be that they represent something categorically different from the other observations. Or, it could be that their values arise due to error. In either case, their presence in the data could distort the analysis, creating misleading results and conclusions.

Figure 2.10 shows there are some counties that have few days when ozone is the main air pollutant (but none with unusually many days). There are actually three counties to the left of the whisker, though two share the same value. That said, when

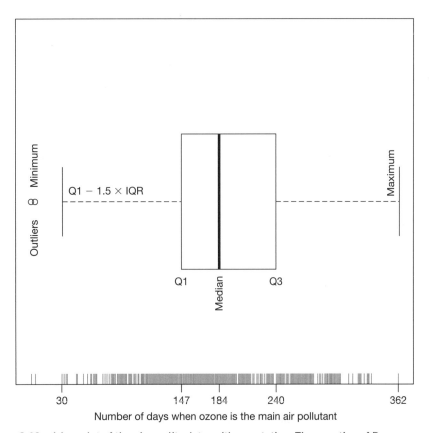

Figure 2.10 A box plot of the air quality data, with annotation. The counties of Bergen, Plumas and Benzie are beyond the whiskers (hinge value of 1.5). They have a low number of days when ozone is the main air pollutant. A rug plot is also shown

39

data display the sort of bell-shaped (humped) distribution suggested by Figure 2.6, we expect about 0.5% of the data values to be beyond each whisker. There are 715 observations; one-half of 1% is 3.575, approximately the same number as we find to the left of the whisker. Looking at the rug plot we find the three counties to be apart but still quite close to the other observations. Our judgement would be to retain them in the data set.

In practice, the '1.5 times' rule can be substituted with any value, so always check how a box plot has been produced. The chosen value is sometimes called the hinge.

Using box plots to compare sets of data

Box plots are useful for comparing variables if they are measured in the same units and have a broadly similar range of values. For example, Figure 2.11 shows that it is far less usual for particulate matter to be the main pollutant in a US county as opposed to ozone. The median for particulates is 11 days with an IQR of 49 days: from 3 to 52 days. However, the particulate data are **skewed** – there are many values that lie beyond

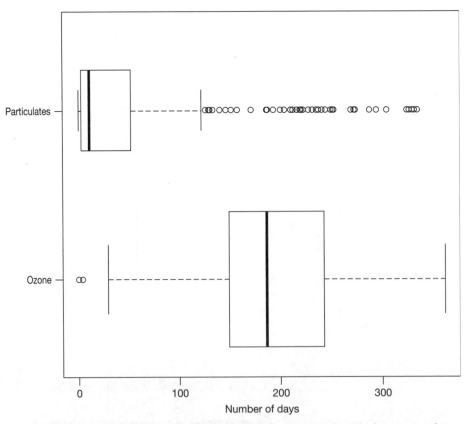

Figure 2.11 Box plots comparing the distributions of the ozone and particulate matter data in selected counties of the United States. The height of the boxes is here proportional to the number of complete observations ($n = 715$ for the ozone data and $n = 425$ for the particulate data)

the right-side whisker. The highest are for Nye County, Nevada (where the Nevada test site for the testing of nuclear weapons is located), and Prowers County, Colorado, at 334 days.

Often the heights of the boxes are arbitrary and simply scaled to the overall size of the graphic. However, they can be drawn proportional to some other values. To demonstrate this, Figure 2.11 is drawn with the boxes sized in accordance with the number of complete observations; that is, removing missing data. The ozone variable has more complete records so is taller.

In general, box plots give a sense of the shape of the data insofar as it is revealed by the median, the IQR, the whiskers and by identifying possible outliers. Further information about the shape of the data can be displayed using a violin plot. Basically, this takes the rectangular shape of the box plot and changes it to the shape of a histogram fitted to the data (Hintze and Nelson 1998; also see the R graph gallery at http://addictedtor.free.fr/graphiques/).

Key concept 2.11 Box plots and the IQR

Box plots are an effective way of summarising the shape of a data set and for comparing variables. They show the median and also the IQR. Whiskers are drawn out from the box and used to identify particularly high or low values.

The five- or six-number summary

A five-number summary is a common way of summarising a variable in a table. This gives the minimum, the first quartile value (Q1), the median (which is also the second quartile, Q2), the third quartile (Q3) and the maximum, and is exactly what a box plot shows. In addition, the mean is often reported, giving a six-number summary. Table 2.8 is a six-number summary of the air quality data with the number of observations also given.

Table 2.8   Summary of the air quality data: number of days in 2007 when ozone was the main air pollutant, as reported for each of 715 US counties, and the number of days when particulate matter was the main pollutant, as reported for 425 US counties

	n	Min.	Q1	Median, $\tilde{x}$	Mean, $\bar{x}$	Q3	Max.
Ozone	715	0	147	184	190	240	362
Particulates	425	0	3	11	44	52	334

Source: the Environmental Agency 2008: http://www.epa.gov/air/data/.

Key concept 2.12 A five- or six-number summary

A **five-number summary** presents the minimum, first quartile, median, third quartile and maximum value of a variable. If the mean is also included it becomes a six-number summary.

2.8 Types of numeric data

We end this chapter as we started, returning to the issue of data types. The categories of nominal, ordinal, interval and ratio data are from the Stevens scales. We also distinguish between counts and rates, between proportions and percentages, and between data measuring additive or multiplicative processes.

Nominal data

Nominal data values are used to indicate categories or classes giving each a numeric code (for example, 1 = USA, 2 = Australia, 3 = Canada, etc.). The categories are usually inclusive and mutually exclusive, meaning that all the objects of study belong to one class (albeit an 'other' or 'unknown' category), and all belong to only one class.

If the values are nominal they are also arbitrary and should not be used to order the classes from highest to lowest, or from most to least. The values are simply labels and not measurements. It therefore makes little sense to calculate any of the six-number summary for nominal data. However, the mode can identify the most frequently occurring class of object and a frequency table can be produced for all classes.

Ordinal data

Ordinal data, like nominal data, may be referred to as categorical data. However, whereas nominal data *just* distinguish one class from the others, ordinal data express a relationship between the categories. The data are now ordered in some way, meaning that it is possible to say that class 1 has more (or less, of something) than class 2, and that class 2 has more (or less) than class 3, and so forth.

Consider the **Likert** scale often used in surveys (Likert 1932). A statement is made (for example, 'I have enjoyed reading this chapter') and respondents indicate whether they 'strongly disagree', 'disagree', 'are neutral', 'agree' or 'strongly agree'. In survey design this is called a closed question because only particular answers are permitted. This allows them to be easily coded using discrete values, for example 1 for strongly disagree, 2 for disagree, 3 for neutral, 4 for agree and 5 for strongly agree. These values are not nominal but ordered: a value of five indicates a greater strength of support than a value of four; a value of four indicates more support than three.

But how much more? We do not know. The values are not fixed to a measurement scale that would permit us to regard the increase between neutrality (3) and agreement (4) to be equivalent to an increase between agreeing (4) and strongly agreeing (5). Neither is a one-unit increase. The values are relative. As a consequence, neither the mean nor the standard should be calculated. Instead, the median and the IQR, the mode and a frequency table can summarise overall levels of support.

Using the Likert scale produces a weakly ordered set of data. It is a weak order because the data are grouped and most or all of the groups will contain many members. The data do not each have a unique value. By contrast, the Index of Multiple Deprivation (IMD 2007) for England and Wales takes the overall deprivation scores assigned to census zones and continuous in the range 85.46 (most deprivation) to 0.37

(least deprivation) and ranks them from 1 (most deprivation) to 32 482 (least). The result is a strongly ordered set of data – the values are not shared.

Interval and ratio data

Interval data are those that can be assigned a position along some scale of measurement. Because of this, the difference between any two values can be measured and quantified.

Ratio data are those derived from a measurement system with a 'natural zero'. Because of this, pairs of values can meaningfully be compared as ratios. For example, 4 km is twice the distance of 2 km, and 1 km is half 2 km. These are ratio data (the ratios are 4:2 and 1:2).

By contrast, it cannot really be said that 50°C is twice as hot as 25°C because, although the base, 0°C, has a physical attribute (the temperature beyond which ice melts), it is not clear what 50°C has double of when compared with 25°C. These are interval data.

Counts and rates

A count measures quantity – how much there is of something. For example, crime data report a total count of 23 156 cars stolen within the West Midlands of England during 2003–2004. For Yorkshire and The Humber the count was less at 19 648 (Office of National Statistics, 2004).

A rate is one quantity divided by another. An obvious example is velocity ('speed') expressed as a measure of the distance travelled, divided by the time taken to do so. Because nearly all government and administrative units differ in physical and population size, rates are used to help compare their attributes and to avoid reliance on raw counts alone. With the car crime data, a reason for the higher number of thefts in the West Midlands is because more people live there. According to the 2001 Census, 1 807 895 people were aged 16–74 and living in the West Midlands, whereas there were 1 489 631 people in Yorkshire and The Humber. If there are more people then there are likely to be more cars and more people to steal them!

It is therefore better to compare the data with respect to the population counts. Dividing the count of car theft by the count of persons gives a measure of the rate of crime in both of the regions. It is 0.012 car thefts per person in the West Midlands, actually slightly lower than the 0.013 car thefts per person in Yorkshire and The Humber.

A rate only makes sense if you divide by a sensible quantity. It is easy to make a mistake. In our example it might be preferable to express the number of car thefts with respect to the total number of cars in each region, or perhaps the number of households, substituting either one of those variables for the number of persons aged 16–74. It depends on the purpose of the analysis and, more pragmatically, the data available to you.

Proportions and percentages

Sometimes, when people talk of rates, they are actually measuring *the share* of a given population that has some characteristic in common. Since the count of those who have it cannot be less than zero and also cannot be greater than the total head count, it follows that the share must always lie within a fixed range of values.

For example, if the unemployment rate is calculated as (a) the count of persons of working age who are fit to work but unemployed, divided by (b) the count of *all* persons of working age who are fit to work (and who may be employed or unemployed), then the result is a proportion that must be in the range from 0 to 1, from nobody to 'the whole lot' being unemployed. The proportion has no measurement units: dividing persons by persons means that the units cancel each other out.

If the proportion is then multiplied by 100, the effect is to stretch the range from 0 to 1 to 0 to 100 instead, giving a percentage (%):

$$\text{proportion, } p = \frac{a}{b} \qquad (0 \le p \le 1)$$

$$\text{percentage, } \% = \frac{a}{b} \times 100 \quad (0 \le \% \le 100)$$

(2.9)

Changing from proportions to percentages makes no difference to the ordering of the data, or to the distribution of high and low values. If you plot the data as proportions and then again as percentages they will look exactly the same as each other on a histogram or in a box plot. Only the measurement scale is different. This is indicated in Equation 2.9 where $(0 \le p \le 1)$ means a proportion has a value within the range from 0 to 1, whereas $(0 \le \% \le 100)$ means the percentage ranges from 0 to 100.

Parametric data

A parametric set of data is one where the shape of the data – its distribution – is or approximates the shape of one of various mathematical models (with known parameters; hence parametric). Approximations can arise because of sampling or measurement errors or it may be that the process under examination may itself not accord with some theoretical ideal.

Nevertheless, having a benchmark can be useful for identifying and correcting errant data and provides a basis for the sorts of probability-based testing which we turn to in the next chapter.

Arithmetic versus other data

A penultimate way to classify data also considers the processes or conditions generating the data. To this point we have been looking at events or phenomena for which a change in their attributes 'adds a little' to the data values. Consider the disease data at the beginning of the chapter. The process of acquiring human rabies is from the bite of an infected animal. However, there is an additional process of medical intervention that means the person is quickly quarantined, preventing the disease from spreading. The net result is that one bite from an infected animal adds the same number of human rabies cases.

Yet diseases often do spread, and quickly, until some intervention occurs or so many of the population are infected that the disease begins to die out. Figure 2.12 shows the number of cattle slaughtered in the UK in each of the months from February to September during the 2001 foot-and-mouth disease outbreak (http://footandmouth. csl.gov.uk/). Foot-and-mouth disease is highly infectious, affecting cloven-hoofed

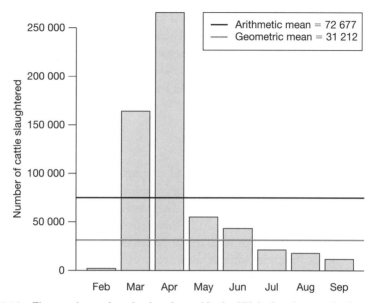

Figure 2.12 The numbers of cattle slaughtered in the UK during the months from February to September 2001, due to foot-and-mouth disease. The arithmetic and geometric means of the data are also shown

Source: http://footandmouth.csl.gov.uk.

animals with fever and blistering. Although most would recover, infected animals are usually culled in the European Union and not vaccinated for economic and welfare reasons (for example, milk yields are affected in cows).

Figure 2.12 suggests a process whereby the disease quickly multiplies and increasingly large numbers of cattle are slaughtered (some because they are infected, others to try and stop further transmission of the disease). Over time, control over the spread of the disease is achieved and the numbers of cattle slaughtered quickly subside.

The result is not a gradual process where the numbers slaughtered gently rise and then fall. Instead, the process is more multiplicative than additive, and the (arithmetic) mean no longer identifies the centre of the data. In this case, the geometric mean is used instead (see Crawley 2005).

Primary and secondary data

A final way to classify data is by how they were acquired. Primary data are those you collect yourself, in the field. They are collected specifically for the research. Secondary data are those that have been collected by someone else, already, for reasons other than your research. They may be obtained from data archives, governmental data reporting and dissemination, from commercial data vendors, and so forth.

In principle, primary data are better insofar as they are task specific and you have control over how they are collected, when and from where. However, primary data collection is expensive, inefficient and of limited scope. Often it makes more sense to share data, especially when so much data are routinely collected for various social, scientific and commercial reasons anyway.

There are important ethical and libertarian questions you could and should ask about using some of the social and personal data that are available (you might read Lyon 2007). However, used sensibly, they add a richness, detail and basis for policy-relevant decision making that would be difficult to achieve from primary data alone (Goodchild and Longley 1999; Longley and Harris 1999).

2.9 Conclusion

In this chapter we have looked at simple techniques to describe numeric data and to present them in tables or using graphics. Such approaches are a good starting point for almost any statistical analysis. They are invaluable for exploring and aiding understanding of the data. However, in themselves they do not identify whether an observation really is unusual or what might be causing the data to vary. To address these issues we need to consider what we would regard as normal, and we do so in the following chapter.

Key points

- When observations are sorted into numeric order, it is easy to find the minimum and maximum values, the median and the first and third quartiles. Together these provide a five-number summary of the data.
- Frequency tables can also be produced, which can then be visualised using bar charts (for a small number of discrete data) or by using histograms (for more continuous data).
- A histogram displays the overall shape of the data – its distribution. Using a rug plot to indicate the individual data values enhances it.
- Measures of central tendency indicate the centre of a set of data. These include the mean, the median and the mode. Their use depends on the data type.
- The mean is susceptible to outliers but is easily calculated, widely used and has useful mathematical properties.
- A measure of central tendency should be accompanied by a measure of spread around it. Possibilities include the interquartile range and the standard deviation.
- The standard deviation is essentially a measure of the average variation of observations around their mean.
- A box plot indicates the centre and spread of one or more variables and is based on a five-number summary of the data.

References

Agresti, A. (2007) *An Introduction to Categorical Data Analysis,* 2nd edn, Hoboken, NJ: Wiley.

Bolstad, W.M. (2007) *Introduction to Bayesian Statistics,* 2nd edn, Hoboken, NJ: Wiley.

Centers for Disease Control and Prevention (2007) Summary of notifiable diseases – United States, 2005. *Morbidity and Mortality Weekly Report,* 54(53).

Crawley, M.J. (2005) *Statistics: An Introduction using R: An Introduction Using R,* Chichester: Wiley.

Dancey, C. and Reidy, J. (2004) *Statistics Without Maths for Psychology,* 3rd edn, Harlow: Prentice Hall.

Field, R. (2003) The handling and presentation of geographical data. In *Key Methods in Geography,* London: Sage, pp.309–341.

Goodchild, M.F. and Longley, P.A. (1999) The future of GIS and spatial analysis. In *Geographical Information Systems: Principles, techniques, management and applications,* New York: Wiley, pp.567–580.

Hintze, J.L. and Nelson, R.D. (1998) Violin plots: a box plot-density trace synergism. *American Statistician,* 52(2), 181–184.

Huff, D. (1954) *How to Lie With Statistics,* New York: W. W. Norton.

Likert, R. (1932) *A Technique for the Measurement of Attitudes,* New York: Columbia University Press.

Longley, P. and Harris, R. (1999) Towards a new digital data infrastructure for urban analysis and modelling. *Environment and Planning B: Planning and Design,* 26(6), 855–878.

Lyon, D. (2007) *Surveillance Studies: An Overview,* Cambridge: Polity Press.

Stevens, S.S. (1946) On the theory of scales of measurement. *Science,* 103, 677–680.

Tufte, E.R. (2001) *The Visual Display of Quantitative Information,* 2nd edn, Cheshire, CT: Graphics Press.

Velleman, P.F. and Wilkinson, L. (1993) Nominal, ordinal, interval, and ratio typologies are misleading. *American Statistician,* 47(1), 65–72.

Wrigley, N. (2002) *Categorical Data Analysis for Geographers and Environmental Scientists,* Caldwell, NJ: Blackburn Press.

The normal curve

Chapter overview

This chapter looks at the normal curve, so called because it is how many social and scientific data appear to be distributed. We consider how the normal distribution arises and find that two key statistics introduced in Chapter 2 – the mean and the standard deviation – are required to define the centre of the curve and the spread of the data around it. We also present ways to detect and to deal with non-normal data.

The focus of the chapter is on understanding the concept of normality, applying it to real-world data. Whilst the chapter is not a treatise on probability theory, the link between statistics and probability is important, providing the basis for determining whether the result of a statistical test is unusual and worthy of our interest. A brief introduction to probability is given, presenting it as a measure of what would happen in the long run if 'an experiment' were repeated over and over again.

Learning objectives

By the end of this chapter you will be able to:

- Explain why the normal curve provides a model of how many data sets are distributed.

- Use graphical techniques to illustrate a normal distribution and to detect the presence of outliers, skew and other signs of non-normality.

- Explore the use of transformations to correct for skewed data.

- Know how to convert data into z values, which are units of standard deviation from the mean.

- Remember that 95% of the area under a normal curve is within 1.96 standard deviations of the mean, and that 99% of the area is within 2.58 standard deviations.

- Understand how the properties of a normal curve are used to form probabilistic statements about the rarity (or otherwise) of a measured value – to assess whether it is unusual given the centre and spread of the data.

3.1 Introducing the normal curve

We begin looking again at the US air quality data from Chapter 2. You will remember that these were obtained from the Environmental Protection Agency's AirData reports (http://www.epa.gov/air/data/) and record the number of days for which ozone was the main air pollutant in each of 715 US counties in 2007. Table 3.1 provides a summary of the data – a six-number summary combined with the standard deviation and the number of observations.

Figure 3.1 shows the shape – the distribution – of the air quality data. The histogram appears humped or bell shaped with more of the observations having values close to the centre and fewer with values further away. To emphasise this, a bell shape has been superimposed upon the data. The curve shown has the same mean and standard deviation as the data.

The proper name for the bell-shaped curve is the **normal curve** or normal distribution. It is also called the Gaussian distribution, named after the German mathematician (Johann) Karl Friedrich Gauss (1777–1855). It was actually discovered by the French mathematician Abraham De Moivre (1667–1754) (see Pearson 1924) and described as normal, meaning commonplace, by the English polymath Sir Francis Galton (1822–1911) (Frank and Althoen 1994, after Kruskal 1978).

Table 3.1 Summary of the air quality data: number of days in 2007 when ozone was the main air pollutant, as reported for each of 715 US counties

n	Min.	Q1	Median, $\tilde{x}$	Mean, $\bar{x}$	Q3	Max.	s
715	0	147	184	190	240	362	66

Source: the Environmental Agency 2008: http://www.epa.gov/air/data/.

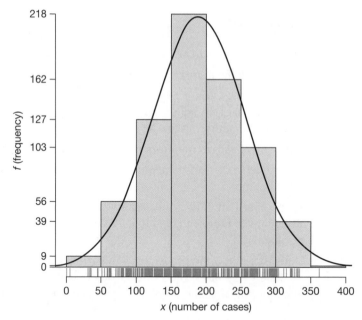

Figure 3.1. A histogram and rug plot showing the distribution of the air quality data: number of days when ozone is the main air pollutant in 715 US counties. A normal ('bell-shaped') distribution is also shown. It has the same mean and standard deviation as the air quality data

Although we may talk of 'the' normal curve, it is actually *a* normal curve shown in Figure 3.1. It is one from the family of such distributions. Other normal curves are shown in Figure 3.2 and it is obvious that they do not look the same. However, not all rectangles, triangles or circles look the same either; some are bigger than others, but we still regard them as having the same shape.

What links the family of normal curves together is that they all have the same form of equation defining their shape. In more sophisticated terminology, they have the same probability density function. If a set of data were perfectly normal their shape would be defined by the equation

$$y = \frac{1}{s\sqrt{2\pi}} e^{-(x-\bar{x})^2/2s^2}$$

(3.1)

This equation may appear daunting but if you look at it closely you will see that it is being driven by two parameters: $\bar{x}$, the mean of the data, and s, the standard deviation. It is these that define where the curve is centred and how spread out around the centre it is.

Looking back at Figure 3.1, we find that the shape of the ozone data is not identical to the normal curve. The differences are partly due to using the histogram, which groups the data into bins. However, the differences go beyond the visual.

For example, whereas the normal curve is perfectly symmetrical around the mean (one side is the mirror image of the other), the distribution of the ozone data is not.

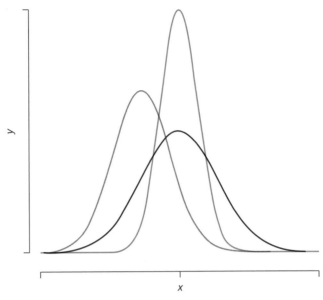

Figure 3.2 A selection of normal curves. They have the same shape but differ in terms of where they are centred and by how much they spread out

Furthermore, the values of the air quality data are restricted in a way that the normal distribution is not. The air quality data are the number of days in 2007 when ozone was the main air pollutant in each of the counties surveyed. That number must lie between 0 and 365, inclusive, and is always a whole number. The normal curve, by contrast, is continuous and actually ranges from negative to positive infinity.

Table 3.2 and Figures 3.3 and 3.4 introduce a new data set with measurements that are genuinely continuous (unlike the quasi-continuous air quality data). These record the straight-line distances that a year group of students travel to get from their homes to a particular elementary school (a primary school) within a British city. The distances were calculated using geographical micro-data provided by the Department for Education (DfE) for research. For brevity, we will refer to the data set as the 'distance data'. The outlier shown in the figures is omitted from the second row of Table 3.2 and from the analysis of the data, which follows.

Table 3.2 Distance travelled to school by a year group of students attending a particular elementary school in a large British city. In the second row of the table, one outlier has been omitted

n	Min.	Q1	Median, $\tilde{x}$	Mean, $\bar{x}$	Q3	Max.	s
59	156.3	508.8	721.1	693.0	848.0	1670.0	295.8
58	156.3	508.8	721.1	676.2	841.3	1257.0	268.3

Source: Authors' own calculations based on data provided by the DfE.

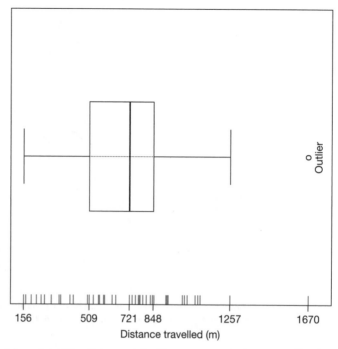

Figure 3.3 A box plot of the distance data. There appears to be one outlier that will be excluded from subsequent analysis

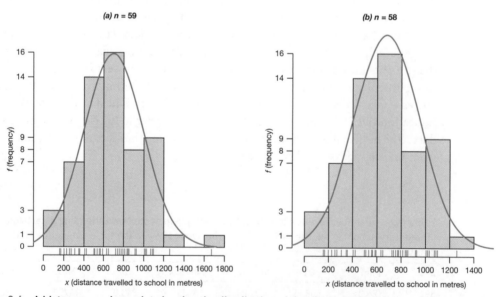

Figure 3.4 A histogram and rug plot showing the distribution of the distance data: distance travelled to school by a year group of students attending a particular elementary school in a British city. A normal curve is fitted to the data (a) and in (b) an outlier is omitted

> ### Key concept 3.1 Normal curves
>
> Many socio-economic, demographic and scientific data sets approximate a **normal distribution**. That is why it is called normal: because it is commonplace!
>
> The normal distribution is sometimes described as bell shaped. If a variable is normally distributed and shown as a histogram then the most frequently occurring values will be found at the centre. The further away from the centre each bin is, the less frequently those values will be found in the data.
>
> To draw a normal curve you need two pieces of information: its centre (the mean) and its spread (the standard deviation).

3.2 Models, error and uncertainty

Looking at Figure 3.4, it is clear that the distance data are not perfectly normal. However, we may treat the normal curve as an abstraction, a model of the distance data.

A definition of **a model** given by the *Paperback Oxford English Dictionary* is 'a simplified mathematical description of a system or process'. The emphasis on the system or process is important. Ultimately, when we model data, what we actually are interested in is not the data as such but the properties of whatever has been measured. In this respect the data are an end product, of interest because they provide a way to understand the system or process under investigation.

The dictionary definition also talks of simplification. The virtue of simplification is presented by Kreft and Leeuw (1998) in a discussion of multilevel modelling (a technique we look at in Chapter 9):

> [S]tatistical models are mathematical models. If the data generation closely resembles the assumptions underlying the statistical model the chances are larger that the conclusions based on the results are close to reality. But reality, in all its complexity, cannot be modeled in a useful way. Complex models may imitate well, but will be equally complex, and thus useless tools. Summarizing data in a complex way is not a step forward.

(p.x)

We could nuance the above and suggest that *over*simplification is not a step forward, either, noting, for example, that climate models predicting climate change are extremely complex. However, the point remains: a statistical model is a simplification, an abstraction of a more complex reality, an aid to understanding.

In our case, the model is a normal curve and has the equation

$$y = \frac{1}{\sigma\sqrt{2\pi}} e^{-(x-\mu)^2/2\sigma^2} \tag{3.2}$$

This is the same as Equation 3.1 except we are now using μ and σ to indicate the centre and standard deviation of the curve. We will talk more about the distinction between $\bar{x}$ and μ, and between s and σ, later in the book. For now it is sufficient to understand $\bar{x}$ and s as properties of the data, whereas μ and σ are properties of the theoretical model. If the difference is unclear, it might be helpful to understand μ and σ as representing the true mean and standard deviation, which we would obtain only if the data were entirely free from error and the model was perfectly correct.

We know that the distance data are not perfectly normal but does that make the normal curve the wrong model to fit to the data? Not necessarily. It is possible that the normal curve provides a better representation of the true shape of the distance data than that provided by the data alone. This amounts to saying that we do not fully trust the data. Why not?

The heart of the issue was explored in Chapter 1. There it was noted that any measurement contains error. It is, we concede, possible that the occasional measurement is perfectly correct but you would not know because you lack a benchmark to validate it. If you know for sure what the true value is, then why bother to measure it? Any measurement is estimation.

In everyday language, an **error** is understood as something that could be avoided. The source of the error is known and may therefore be corrected. Here, however, errors are understood not just as mistakes at the point of measurement but from all things that impact on the process of measurement, that affect the data values and which create misreadings, no matter how slight, of the actual state or condition of the feature, system or process being studied.

Some imperfections will come from the measurement device used, some will arise from how the data are collected (see Chapter 4 about sampling) and others will arise from the complexity of the system being studied and all the possible factors that affect the observations. An umbrella term for all these 'things' is sources of **uncertainty**. There are many sources of uncertainty present in the social and geographical sciences, because people, places, socio-economic structures and physical environments are complex, entangled and difficult to contain in a laboratory!

In general, the combined effects of these various and often unknowable errors are assumed to be random upon the data. That is, they are assumed not to change systematically or shift fundamentally our view of what is being studied. Instead, they act 'in all directions', blurring the clarity of what is seen but still leaving us looking in the right direction. To return to the analogy of the radio signal begun in Chapter 1, the errors cause crackle in the signal but do not leave us tuned to the wrong radio show. In sampling terms, the **random errors** are sources of imprecision but not of bias(/inaccuracy).

It is important to retain a sense of perspective. To say that data are not perfect is not to say they are completely imperfect or entirely unhelpful. The truth lies somewhere in between. Returning to the distance data, their shape gives reason to suppose that they might be normally distributed, or at least that the normal curve is an adequate model of the shape of the data if all sources of measurement error could be taken away.

Here there is interplay between data and theory, with the data either suggesting or validating a theoretical model that can then be used as an alternative to the data, in preference to the data, or as a basis for comparison with the data. In this case we will use the normal curve.

> **Key concept 3.2 Modelling data using the normal curve**
>
> Many socio-economic, demographic and scientific processes give rise to data that appear normally shaped. However, those processes are never measured perfectly and without error, and those imperfections will impact on how the data are distributed. The normal curve is one of a number of distributions that provide a **theoretical model** of the true shape of the data or an approximation for it.

3.3 Why are the data (approximately) normal?

If we are satisfied that the distance data are normally distributed (or approximately so) then a reasonable question is 'why?' Why do many social and scientific data sets have a shape that is common enough to be called normal?

For an explanation, we will borrow part of a data set complied by Albert (2007) (available in R's LearnBayes package: http://cran.r-project.org/). It lists the height of 647 university students (from 657) who gave their height in answer to a survey question. Each height is recorded to the nearest whole inch in the data set. The data are summarised by Table 3.3 and their distribution by Figure 3.5.

Like the distance data, the height data are not perfectly normal but display properties of normality. In particular, students with values close to the mean height (of 66.7 inches) are more frequently observed in the sample than those with relatively short or tall heights. The interpretation of the data is intuitive: more students are of close to average height than are either (very) short or tall. But why?

For an explanation, consider what causes an adult's height. There is a genetic component that is inherited from the parents (and from their parents, and so on through the generations). The genetic factors are complicated. Will a tall adult choose another tall adult or is sexual attraction more varied, or constrained by social circumstances? Does tallness always beget tallness or is there more of a randomness (or complexity) to our gene pool?

What is more, height is determined by other factors, including diet, health and lifestyle such as smoking. According to Silventoinen (2003), about 20% of variation in body height is due to the environment in which children are raised, with nutrition and disease being the main reasons why.

Assume that there are multiple genes affecting the height of an individual. Assume also that there is additional variation caused by other factors, some of which are not known. To be very tall (or very short) requires all those factors, including the genes,

Table 3.3 Summary of the student height data: height of survey respondents self-reported in inches. The NA (Not Applicable) records the number of missing data. In this example, 10 students did not report their height

n	Min.	Q1	Median, $\tilde{x}$	Mean, $\bar{x}$	Q3	Max.	s	NA
647	54	64	66	66.7	70	84	4.3	10

Source: Albert (2007).

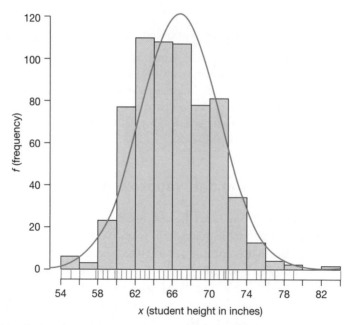

Figure 3.5 Distribution of the student height data presented in Albert (2007). A normal curve is fitted to the data

to act together in a particular direction: to cause an increase (or a decrease) around an otherwise expected height.

Intuitively, having *all* the factors acting in one direction is less likely than *many* of the factors doing so, which is less likely than *some* acting to increase (or decrease) the height of the person concerned. In this way, exceptionally tall people are rarer than very tall people, very tall people are less common than quite tall people, and quite tall people are less frequently observed than tall-ish people. The same is true of short people.

The intuition helps explain Galton's (1886) observation that children of very tall parents, whilst themselves tall, tend also to be shorter than their parents. Put crudely, the parents got lucky on the factors governing height. It requires quite exceptional luck for the parents *and* the children to be very tall.

What Galton observed is described as **regression towards the mean**, a long-run movement towards average height in the children of very tall or short parents. The situation is akin to tossing lots of coins and assessing the chances of all, most or some landing head up. Some is more likely than most, and most is more likely than all doing so. A difference is that each coin contributes equally to the chance of getting a head or not, whereas the factors governing height are unlikely to be equal in their effect. However, as Grinstead and Snell (2006, p.345) note, '[essentially] anything that can be thought of as being made up as the sum of many small independent pieces [factors, causes or coins] is approximately normally distributed'.

The statement above is a generalisation of what is known as the **central limit theorem** and is the key to understanding why many socio-economic, demographic and scientific data sets are normally distributed (or approximately so). The theorem can be applied to the distance data. Undoubtedly there are many behavioural, social and physical factors that govern how far a pupil travels to their school. If the net result of these can be regarded as 'the sum of many small independent factors' then the data ought to be as they are, (approximately) normally distributed, provided that the size of the data set is not too small (a condition we discuss further in Chapter 5).

Key concept 3.3 The central limit theorem

The **central limit theorem** states that the sum (and average) of a large number of independent and identically distributed random variables will approach a normal distribution as the sample size increases, even if those variables are not themselves normally distributed.

A **random variable** is a set of measurements, the values of which are a combination of the true signal and of random noise. Because random events affect each measurement, no value can be predicted in advance with absolute certainty, and no value is definitive. Instead, possible measurements have a probability of occurring.

Nearly all variables used in statistical analysis have a random part. If the data you are analysing are the measured outcomes of many underlying and broadly independent factors (not all of which you will necessarily know about) then the central limit theorem says those measurements will be normally distributed, or approximately so.

3.4 Important properties of a normal curve

A normal curve ranges from negative to positive infinity, with a continuous set of values in between. The distribution is symmetrical around its centre, the mean, with more of the data being close in value to the mean than far from it, giving the classic bell shape. In practice, a variable will tend to have an upper or lower limit – heights, for example, cannot be negative – and the measured values will not be truly continuous, just more continuous than they are discrete.

An important property of any truly normal distribution is that 95% of the area under the curve is within a set distance either side of the mean. That distance is equal to 1.96 multiplied by the standard deviation. It is also true that 99% of the area under the curve is within a distance equal to 2.58 multiplied by the standard deviation. Figure 3.6 illustrates these properties of a normal curve where the letter z is used to indicate the number of standard deviations from the mean.

These properties of the normal curve are so often used in statistical work that they are worth memorising: 95% of the area within 1.96 standard deviations of the mean; 99% within 2.58 standard deviations.

Key concept 3.4 Properties of a normal curve

Any normal curve is symmetrical around its mean. Half the area is found to the left of the mean and the other half to the right.

Of statistical importance is the fact that 95% of the area under any normal curve is found within 1.96 standard deviations either side of the mean, and 99% within 2.58 standard deviations.

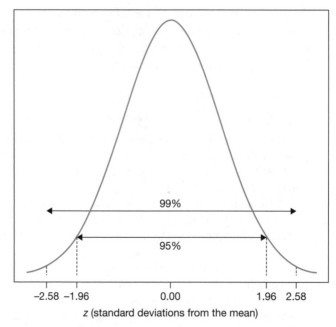

Figure 3.6 Showing that 95% of the area under a normal curve is within 1.96 standard deviations of the mean and that 99% of the area is within 2.58 standard deviations

3.5 An experiment

Looking back at Figure 3.4, our model of the distance variable is a normal distribution with a mean of $x = 676.2$ and a standard deviation of $s = 268.3$ metres (the same as in the second row of Table 3.2).

As we know, 95% of the area under a normal distribution is within 1.96 standard deviations of the mean. For the distance data, the value 1.96 standard deviations below the mean is

$$\bar{x} - 1.96s = 676.2 - \left(1.96 \times 268.3\right)$$
$$= 676.2 - 525.9$$
$$= 150.3 \tag{3.3}$$

The value 1.96 standard deviations above the mean is

$$\bar{x} + 1.96s = 676.2 + \left(1.96 \times 268.3\right)$$
$$= 676.2 + 525.9$$
$$= 1202.1 \tag{3.4}$$

Therefore, 95% of the area under the normal curve is between the values of 150.3 and 1202.1.

We now take a single value at random from the normal curve, then two values, three, four . . . and so on until we have 10 000 values and 10 000 sets of data, each set having

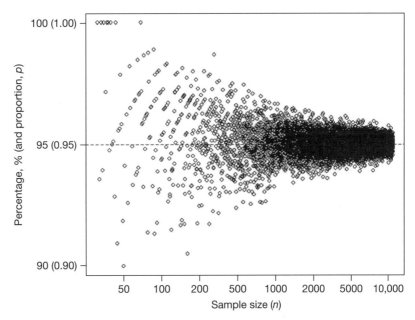

Figure 3.7 The percentage (and proportion) of values that are within 1.96 standard deviations of the mean for each of 10 000 batches of data randomly drawn from a normal curve. Note that as the size of the batch increases, the percentage tends towards 95%. (The percentages actually ranged from 0 to 100% but sample sizes less than $n = 30$ and the occasional value less than 90% are not shown)

one more member than the set that preceded it, and with each set having values that are drawn independently of each other (so one set can contain entirely different values to the next).

Finally, we take each set in turn and calculate the percentage of its values that are within 1.96 standard deviations of the mean. The results are shown in Figure 3.7. Sometimes the percentage is higher, sometimes it is lower. However, as the sample size increases, the percentages converge on the 95% value.

With the above in mind, let us pose you a question. Assume that our model of distance travelled from home to school is representative of all pupils in the school. We now select 100 of those pupils at random. How frequently would you expect values between 150.3 and 1202.1 metres to be the distances those pupils travel?

Your answer:

Based on the model, you should expect the frequency to be correct 95 times: 95 in 100, which is 95%. This is because 150.3 and 1202.1 metres are the distances at 1.96 standard deviations of the mean and 95% of the normal curve is within 1.96 standard deviations of the mean.

In a similar way, if we sampled 100 students from the height data, how many would you expect to have a height of between 55.6 and 77.8 inches? A clue: assume the height variable is normally distributed with a mean of 66.7 and a standard deviation of 4.3 inches. The values 55.6 and 77.8 are at 2.58 standard deviations either side of the mean, that is at $66.7 - (2.58 \times 4.3)$ and at $66.7 + (2.58 \times 4.3)$. The answer is not 95 students but is ... ?

Your answer:

3.6 Probability and the normal curve

In the previous section we imagined taking 100 values at random from a normal curve and asked you to predict the number lying within 1.96 standard deviations of the mean. The answer was 95 of the 100, or 95%. (We also asked you to predict the number within 2.58 standard deviations. The answer was 99 of the 100, 99%.)

The prediction is based on what happens in the long run. As we draw more and more values from the normal curve, the percentage within 1.96 standard deviations of the mean gets closer and closer to 95%. This is what Figure 3.7 is showing; notice how the percentages funnel in toward 95% as the sample size increases from left to right along the horizontal axis.

Taken alone, there is no guarantee that any individual value will be within 1.96 standard deviations of the mean. Some will be, some will not. However, from what we know about the normal curve and what will happen in the long run, we can say that any value randomly drawn from a normal curve has a 95-in-100 chance of being within 1.96 standard deviations of the mean. In the same way, we can say that there is a 99-in-100 chance that the randomly drawn value will be within 2.58 standard deviations of the mean.

We can express these values as **probabilities**:

$$P(\mu - 1.96\sigma < x < \mu + 1.96\sigma) = 0.95 \tag{3.5}$$

$$P(\mu - 2.58\sigma < x < \mu + 2.58\sigma) = 0.99 \tag{3.6}$$

Do not be put off by the Greek symbols! Equation 3.5 simply is saying that the probability of the value (x) being within 1.96 standard deviations (σ) of the mean (μ) is 0.95. Equation 3.6 is saying that the probability of its being within 2.58 standard deviations is 0.99.

Previously, in Section 3.2, we alluded to when we use μ to indicate the mean and σ for the standard deviation. We elaborate a little more here. The normal curve is 'the thing' (called the population) from which a sample of values is taken. Its mean is represent by μ and its standard deviation by σ. Both of these parameters are assumed to be fixed and unchanging.

In contrast, the mean and standard deviation of a sample data that is taken from the population are represented by $\bar{x}$ and s. Their values depend on whatever data have been sampled and are therefore variable, changing from one set of data to the next. This helps explain why it is important to distinguish properties of the sample from properties of the population from which the sample has been drawn.

Probability expresses the likelihood that a particular outcome (or sequence of outcomes) will arise from an experiment, trial or test. A probability of 0 means an outcome will definitely *not* occur regardless of how many times the experiment is repeated. A probability of 1 means the outcome will *always* occur every time the experiment is undertaken.

Midway between the extremes, a probability of 0.5 means the outcome is as likely to occur as not. A probability greater than 0.5 but less than 1 means the outcome is *more* likely to occur than not but will not definitely happen. Conversely, a probability of less than 0.5 but greater than 0 means the outcome is *less* likely to occur than not but could happen anyway.

If the experiment could be repeated many times then, in the long run, an outcome has a probability of 0.95 if it is expected to occur in 95% of the experiments. Expressed as a proportion, this is $p = 0.95$. Referring to each experiment as a trial, and the number of times a particular outcome occurs as 'a success', then the frequency of success relative to the number of trials is expected to be 95-in-100.

In general, if p is the proportion of the trials expected to be successful – the probability of success – and n is the number of trials, then the *predicted* frequency of success is

$$\hat{f} = n \times p \tag{3.7}$$

where 'the hat' on the f indicates the frequency of success we *expect* but may not actually *observe*. (Try repeatedly tossing a coin 10 times: does it always land head up for exactly half the tosses?) If $p = 0.95$ and there are 100 trials, 95 of them are expected to be successful. If there are 200 trials, 190 of them are expected to be successful.

The frequency of success relative to the number of trials can be written as the expression

$$\frac{f}{n} \tag{3.8}$$

What Figure 3.7 has shown is that as the number of trials increases towards infinity, this proportion moves towards the true probability of success:

$$\frac{f}{n} \to p \quad \text{as} \quad n \to \infty \tag{3.9}$$

A probability can therefore be regarded as measuring the frequency of success relative to a large number of trials. That is, it expresses what would happen in the long run.

There are four caveats. First, it is assumed that the trials are fair and not biased to a particular outcome. Secondly, it is assumed that the results of the trials are independent of each other. For example, tossing a coin and getting a head a first time does not alter the probability of getting a head (or a tail) the next time (both remain constant at $p = 0.5$).

Thirdly, it is assumed that the system being studied is stable and not changing appreciably. This is true of a coin where nothing appreciable changes each time it is tossed, but what does it mean to measure a long-run frequency for some aspect of a fast-changing social or physical system (for example, the probability of a flood event in a changing climate)?

Fourthly (and not unrelated to the previous point), the long-run frequency of an outcome occurring is not the only understanding of probability used in statistical research. Another is to regard probability as quantifying a degree of belief in something, given the evidence for it. The latter is a Bayesian perspective.

An advantage of the Bayesian perspective is that it relaxes the assumption of the population mean being fixed and only truly determinable by an infinite number of data collections. Instead, it may be given a probability density function, meaning some values are more plausible than others and our information about which are so can be based on any prior expectations we have, updated for whatever the data are saying.

The techniques we present in this book are frequentist: they estimate the frequency of an outcome occurring in the long run. However, they do have Bayesian counterparts (see Bolstad 2007). Bayesian techniques are used in remote sensing to help match the spectral signature of what is observed to the type of land use it represents (Mesev 2003).

Key concept 3.5 Probability and the normal curve

The probability that an experiment or trial generates a particular outcome can be estimated by repeating the experiment many times and calculating the relative frequency of success. This is f/n where f is the number of successful outcomes and n is the number of trials.

As the number of trials increases, the relative frequency moves closer to the true probability of success. In this way, **a probability measures the frequency of success relative to a large number of fair and independent trials**. By fair it is meant that the trials are not biased in a way that distorts the true outcome. By independent it is meant that the outcome of one trial does not affect the outcome of other trials.

Drawing a value at random from a normal curve can be regarded as a trial where the probability of successfully drawing a value within 1.96 standard deviations of the mean is $p = 0.95$. The probability of successfully drawing a value within 2.58 standard deviations of the mean is $p = 0.99$.

3.7 Worked examples

Table 3.4 summarises the amount claimed by Members of Parliament in the UK as expenses in the year 2006–2007 (before the 'expenses scandal' of 2009). A box plot of the data is shown in Figure 3.8. Figure 3.9 shows the distribution of those data having excluded three of the more frugal MPs identified in Figure 3.8.

Example 1

Assume the expenses data are a normal variable with a mean of £136 181 (that is, about five times the average household income in the UK) and a standard deviation

Table 3.4 Amount in pounds sterling (£) claimed by Members of Parliament in the UK in the year 2006–2007. In the second row of the table, three outliers have been omitted

n	Min.	Q1	Median, $\tilde{x}$	Mean, $\bar{x}$	Q3	Max.	s
645	44 551	127 843	137 781	135 850	145 687	185 421	15 408
642	91 737	127 940	137 836	136 181	145 718	185 421	14 620

Source: House of Commons.

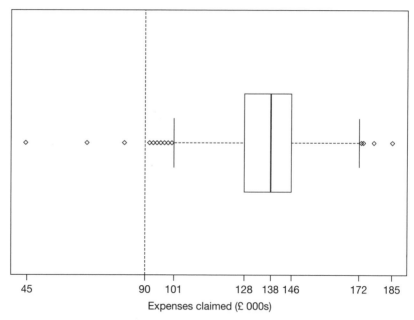

Figure 3.8 Box plot of the expenses data: amount claimed by each of 645 MPs in the UK for the year 2006–2007. Three more frugal MPs are identified amongst the possible outliers (they are to the left of the dashed line)

Source: House of Commons.

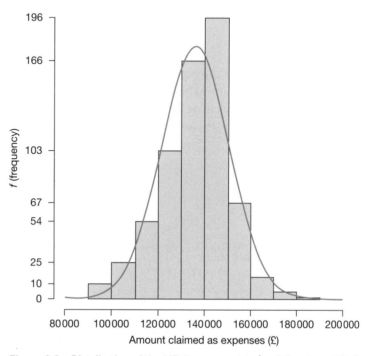

Figure 3.9 Distribution of the MPs' expense data (omitting three MPs)

of £14 620. If we select an MP at random from the data (excluding the three outliers) what is the probability that the MP will have claimed between £107 526 and £164 836 in expenses?

To answer the question, we first calculate how many standard deviations each of the two values is from the mean. For the lower value:

$$z_{LOW} = \frac{x_{LOW} - \bar{x}}{s}$$

$$= \frac{107\,526 - 136\,181}{14\,620}$$

$$= -1.96 \tag{3.10}$$

For the higher value:

$$z_{HIGH} = \frac{x_{HIGH} - \bar{x}}{s}$$

$$= \frac{164\,836 - 136\,181}{14\,620}$$

$$= 1.96 \tag{3.11}$$

The values of 107 526 and 164 836 are at 1.96 standard deviations either side of the mean. The proportion of the area under a normal curve that is within 1.96 standard deviations of the mean is $p = 0.95$ (95%), which is the probability that the MP will have claimed between £107 526 and £164 836 in expenses. Using notation,

$$P(107\,526 < x < 164\,836) = 0.95 \tag{3.12}$$

Example 2

Still assuming that the expenses data are normal, what is the probability of selecting an MP who has claimed between £98 461 and £173 901?

As before, we calculate the number of standard deviations each of the two values is from the mean. For the lower value:

$$z_{LOW} = \frac{x_{LOW} - \bar{x}}{s}$$

$$= \frac{98\,461 - 136\,181}{14\,620}$$

$$= -2.58 \tag{3.13}$$

For the higher value:

$$z_{HIGH} = \frac{x_{HIGH} - \bar{x}}{s}$$

$$= \frac{173\,901 - 136\,181}{14\,620}$$

$$= 2.58 \tag{3.14}$$

The values are each at 2.58 standard deviations either side of the mean. The proportion of the normal curve within 2.58 standard deviations of the mean is $p = 0.99$ (99%), which is the probability that the MP will have claimed between £98 461 and £173 901 in expenses:

$$P(98\,461 < x < 173\,901) = 0.99 \tag{3.15}$$

3.8 *z* values and the standard normal curve

In the worked examples above, we chose values that were at 1.96 or 2.58 standard deviations from the mean of a normal curve. It would be very limiting if we could only assign probabilities to such values. What we need is a more general methodology.

Imagine we were interested in determining the probability of selecting an MP at random who had claimed between £103 287 and £159 134. As before, we calculate how many standard deviations each value is from the mean of the data.

For the lower value:

$$z_{\text{LOW}} = \frac{x_{\text{LOW}} - \overline{x}}{s}$$
$$= \frac{103\,287 - 136\,181}{14\,620}$$
$$= -2.25 \tag{3.16}$$

For the higher value:

$$z_{\text{HIGH}} = \frac{x_{\text{HIGH}} - \overline{x}}{s}$$
$$= \frac{159\,134 - 136\,181}{14\,620}$$
$$= 1.57 \tag{3.17}$$

What we are doing here is converting the data into **z values** – units of standard deviation from the mean. A negative *z* value indicates that the original value is less than the mean. A positive value indicates it is greater. Standardising the data allows them to be modelled using a standard normal curve. This is a special example of a normal curve with a mean of zero and a standard deviation of one. The total area under the standard normal curve is also equal to one.

Standardising the data involves a two-stage process of moving then squashing the data. The first stage is like taking a ruler to measure the length of something but then moving the ruler so it is positioned at the midway point of that something. More specifically, the zero is positioned at the centre – the mean – of the data set. This is what happens when we subtract the mean from the original value (as in the upper parts of Equations 3.16 and 3.17).

The second stage changes the scale of the measurement units by dividing the original units by the standard deviation of the data set, *s*. In this way, the *z* value expresses how far an initial value is from the mean of the data, relative to the spread of it. This

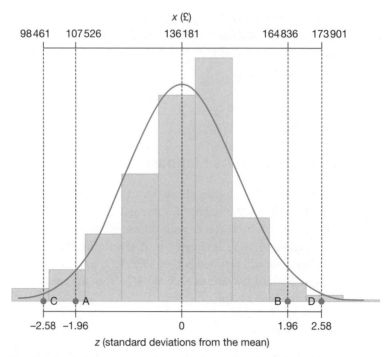

Figure 3.10 Standardising a data set does not change its shape, just the measurement scale. A z value gives the number of standard deviations an observation is from the mean for the data

allows data sets to be compared even when their means and standard deviations are different.

Standardising the data set does not change its shape. If the distribution was normal to begin with, it will still be normal afterwards. And if it was not normal before, then it will not be afterwards, either.

It also does not change the position of any observation relative to the others. In Figure 3.10, observation A remains at 1.96 standard deviations less than the mean regardless of whether the data are expressed using the initial data or standardised into z values. Similarly, observation B remains at 1.96 standard deviations greater than the mean in either case. Observation C remains at 2.58 standard deviations less than the mean, and observation D at 2.58 standard deviations greater. Relative to each other, the observations do not move. It is only the measurement scale that is changing. The principle is really no different from converting counts into percentages, or percentages into proportions.

Next, we determine what proportion of the area under a standard normal curve is between the lower and higher z values. In our example we are interested in the area between $z = -2.25$ and $z = 1.57$. This is shown in Figure 3.11.

To find out, we use a statistical table (or, less anachronistically, a computer-based equivalent). There are various ways a statistical table can report the area, so always check carefully what information you are being given. One of the more common is to give the area under the curve from the far left of the horizontal axis (where $z = -\infty$)

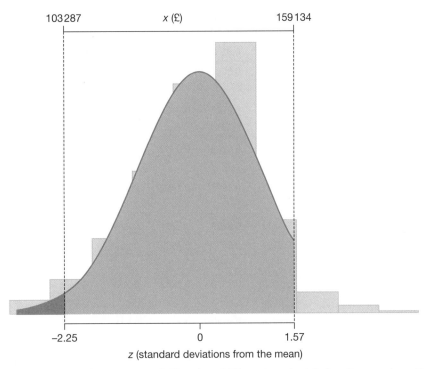

Figure 3.11 The area between *z* = −2.25 and *z* = 1.57 can be calculated as the area from the left to *z* = 1.57 *minus* the area from the left to *z* = −2.25

to specific values of *z*. We will call this 'the left to right method' and is the one we will adopt.

To find the area between *z* = −2.25 and *z* = 1.57 we need to combine the left to right method with subtraction. What we have to calculate is the area from the far left to *z* = 1.57 *minus* the area from the far left to *z* = −2.25. What remains is what is needed and is the larger shaded area in Figure 3.11.

Expressed using notation, what we are seeking is $P(-2.25 < z < 1.57)$ where

$$P(-2.25 < z < 1.57) = P(z < 1.57) - P(z < -2.25) \tag{3.18}$$

Or, more generally,

$$P(z_{\text{LOW}} < z < z_{\text{HIGH}}) = P(z < z_{\text{HIGH}}) - P(z < z_{\text{LOW}}) \tag{3.19}$$

What we now need is a statistical table giving the values from the left of the standard normal to particular values of *z*. Part of one is shown in Table 3.5. The row names give a *z* value to one decimal place. The columns give the second decimal place. Highlighted in the table are $P(z < 1.57) = 0.9418$ and $P(z < -2.25) = 0.0122$. Using the 'left to right and minus' way of working,

$$P(-2.25 < z < 1.57) = P(z < 1.57) - P(z < -2.25)$$

$$= 0.9418 - 0.0122$$

$$= 0.9296$$

$$\approx 0.93 \tag{3.20}$$

Table 3.5 Part of a statistical table showing the area under a standard normal curve to selected values of z. Highlighted are $P(z < -2.25) = 0.0122$ and $P(z < 1.57) = 0.9418$

z	0.00	0.01	0.02	0.03	0.04	0.05	0.06	0.07	0.08	0.09
−2.5	0.0062	0.0060	0.0059	0.0057	0.0055	0.0054	0.0052	0.0051	0.0049	0.0048
. . .										
−2.2	0.0139	0.0136	0.0132	0.0129	0.0125	0.0122	0.0119	0.0116	0.0113	0.0110
. . .										
−1.9	0.0287	0.0281	0.0274	0.0268	0.0262	0.0256	0.0250	0.0244	0.0239	0.0233
. . .										
−1.0	0.1587	0.1562	0.1539	0.1515	0.1492	0.1469	0.1446	0.1423	0.1401	0.1379
. . .										
0.9	0.8159	0.8186	0.8212	0.8238	0.8264	0.8289	0.8315	0.8340	0.8365	0.8389
1.0	0.8413	0.8438	0.8461	0.8485	0.8508	0.8531	0.8554	0.8577	0.8599	0.8621
. . .										
1.5	0.9332	0.9345	0.9357	0.9370	0.9382	0.9394	0.9406	0.9418	0.9430	0.9441
. . .										
1.9	0.9713	0.9719	0.9726	0.9732	0.9738	0.9744	0.9750	0.9756	0.9762	0.9767
. . .										
2.5	0.9938	0.9940	0.9941	0.9943	0.9945	0.9946	0.9948	0.9949	0.9951	0.9952
2.6	0.9953	0.9955	0.9956	0.9957	0.9959	0.9960	0.9961	0.9962	0.9963	0.9964

Source: Authors' own calculations using R version 2.6.1.

This gives us the answer we were looking for. Rounding to two decimal places,

$$P(-2.25 < z < 1.57) = 0.93$$
$$\Rightarrow P(103\,287 < x < 159\,134) = 0.93$$

(3.21)

The probability of drawing a value of between £103 287 and £159 134 from the expenses data is estimated to be $p = 0.93$.

Key concept 3.6 Converting data into z values

Data are **standardised** if instead of their original measurement units they are converted into units of standard deviation from the mean. The resulting values are often called **z values** and are calculated using the formula

$$z = \frac{(x - \bar{x})}{s}$$

where x is the value to be standardised, $\bar{x}$ is the mean of the data set and s is the standard deviation.

A z value expresses how far a value is from the centre, relative to the spread of the data.

Example 3

Assuming that the expenses data are normal with a mean of £136 181 and a standard deviation of £14 620, what is the probability of selecting, at random, an MP who has claimed between £150 000 and £175 000 in expenses?

First, we convert the two values into z values:

$$z_{\text{LOW}} = \frac{150\,000 - 136\,181}{14\,620}$$

$$= 0.95 \tag{3.22}$$

$$z_{\text{HIGH}} = \frac{175\,000 - 136\,181}{14\,620}$$

$$= 2.66 \tag{3.23}$$

Secondly, we use Table 3.5 and the 'left to right and minus' method to find the lighter shaded area in Figure 3.12:

$$P(0.95 < z < 2.66) = P(z < 2.66) - P(z < 0.95)$$

$$= 0.9961 - 0.8289$$

$$= 0.1672$$

$$\approx 0.17 \tag{3.24}$$

Rounding to two decimal places, we conclude that $P(150\,000 < x < 175\,000) = 0.17$.

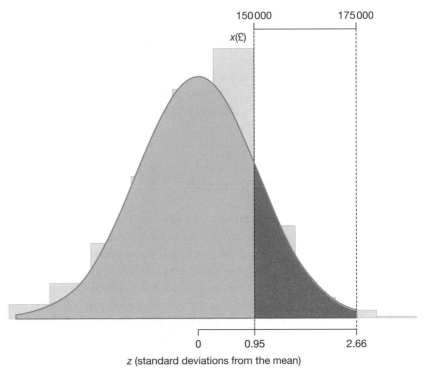

Figure 3.12 The area under the standard normal curve required to give P (150 000 $< x <$ 175 000) for the expenses data

Table 3.6 Summary of the wind velocity data measured in metres per second

n	Min.	Q1	Median, $\tilde{x}$	Mean, $\bar{x}$	Q3	Max.	s
1000	4.058	5.068	5.367	5.384	5.723	6.754	0.471

Source: Data are courtesy of Dr Chris Keylock, School of Geography, University of Leeds.

Example 4

This is an exercise for you to complete by yourself! Use Table 3.5 to confirm that the area under a normal curve that is within 1.96 standard deviations of the mean is $p = 0.95$. A clue: $P(-1.96 < z < 1.96) = P(z < 1.96) - P(z < -1.96)$.

Also confirm that $p = 0.99$ of the area is within 2.58 standard deviations of the mean.

Example 5

Table 3.6 summarises a sample of measurements observing longitudinal velocity within a wind tunnel. The data have been sampled so that the complicating effects of turbulence are removed. Assuming the data are normally distributed, what is the probability that a further measurement within the wind tunnel would yield an answer of between 4.913 and 5.855 metres per second ($m\,s^{-1}$). The answer is $p = 0.68$. Can you use the summary table, Equation 4.14 and Table 4.4 to determine this yourself?

Key concept 3.7   Finding the area under a normal curve

The area between two values under a normal curve estimates the probability of obtaining a measurement that is between those values from a normally distributed (random) variable.

Finding the area involves three stages. First, the mean and the standard deviation of the normal curve need to be estimated from the data. Secondly, z values are calculated. Thirdly, the area between the z values is calculated by using a statistical table or equivalent.

3.9 Signs of non-normality

Although many social and economic data appear normally distributed, not all are. For example, the human rabies data in Chapter 2 are not because they cannot be less than zero, are always whole numbers and do not exceed seven because cases of the diseases are so rare. The shape of these data is better represented by what is called the Poisson distribution and is used to estimate the probability of an event occurring a particular number of times given that the event happens rarely but has considerable opportunity to happen – events such as vehicle collisions at a major road junction.

Data on waves, for example tidal sea levels, have a sinusoidal distribution, rising then falling repeatedly. Other processes display an exponential growth: the bigger 'it' gets, the faster it grows because the effects are multiplicative, not additive (see Chapter 2, Section 2.8). The Malthusian model of population growth is a classic example of this, as is the compound effects of interest on an outstanding debt (the way the debt quickly increases).

Some data have a uniform distribution. Their values are not clustered at the centre but are found with equal frequency across the range. Their distribution is flat – horizontal instead of bell shaped. In a geographical setting, a uniform distribution is when a natural or social entity is evenly distributed across a study region. Examples include creosote bushes in a desert, gulls on a cliff (Avila 1995) and cities in a plain (Glass and Tobler 1971).

Skewed data

Data are often distributed in a way that is similar to normal but with a tail of relatively high or low values. This is especially common for social data, including census data, for which a tail of high values may be found.

As an example, Figure 3.13 shows average weekly income data (2004–2005) for census zones in the UK (Office of National Statistics, 2009). The data are modelled estimates using a combination of survey, census and administrative data. Of interest here is the tail of higher income values to the right of the histogram and of the box plot.

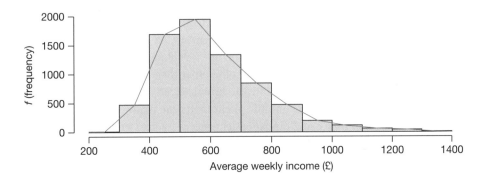

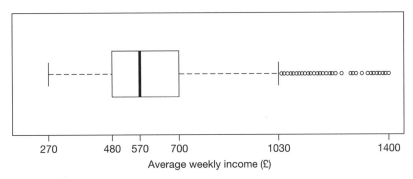

Figure 3.13 The distribution of the income data. There is a tail of higher values, indicating a positive skew. The skew value is Sk = 1.02

The income data are said to be positively skewed, with a tail of values above (to the right of) the mean in Figure 3.13. If a tail of low values is found the data are said to be negatively skewed.

One check on skew is to compare the median and the mean for the data. If the mean is greater than the median than this suggests a tail of high values pulling the mean upwards away from the middle. If the mean is less than the median then there may be a tail of low values pulling the mean downwards.

A more formal way of quantifying skew is to calculate

$$\text{Sk} = \frac{\sum_{i=1}^{n}(x_i - \bar{x})^3}{(n-1)\,s^3} \tag{3.25}$$

where the abbreviation Sk denotes a measure of skewness, $\bar{x}$ is the mean of the data set, s is the standard deviation and $(n-1)$ is the number of degrees of freedom (see Chapter 2, Key concept 2.9). The skewness value for a normal or any other symmetrical distribution is zero. A positive value for Sk indicates a positive skew; a negative value indicates a negative skew. The value for the income data is Sk = 1.02.

Social data are often positively skewed because their values cannot drop below zero – true of the unemployment rate, for example. Admittedly it cannot be above 100%, either. However, the overall unemployment rate in the UK is, at the time of writing, 7.8%, which is much nearer 0% than it is 100%. Consequently, a histogram of regional unemployment rates, for example, could only extend to 7.8 percentage points below the national average but could (in principle) be up to 94.8 points above. When the data are more tightly bounded on one side than on the other then a tail is likely to occur.

A second explanation for how skewed data arise comes by questioning whether some social and economic phenomena can really 'be thought of as being made up as *the sum* of many small independent pieces [factors or causes]'. This, it will be recalled, was the basis for the central limit theorem and for normally shaped distributions. However, it may well be that the factors that determine income – for example, factors such as education, location, opportunity, experience, etc. – are multiplicative rather than additive in their effects. When we apply this thinking to places, it may be that there is an agglomerative effect of 'positive' feedback whereby a more affluent neighbourhood becomes even more exclusive as the effects of housing and labour markets (etc.) price out other people. Such a process would contribute to the growing social and economic inequality that has been observed in the UK over the last 15 years (Dorling *et al.* 2007).

Transforming skewed data

If data are skewed then it may help to transform them in the ways shown in Table 3.7. These are from Erickson and Nosanchuk (1992), after Mosteller and Tukey (1977), and together form 'the ladder of transformation'. Each has the effect of spreading out the lower values and pulling together the higher values (or vice versa, depending on the need), but the strength of their effect differs.

The best transformation for the data can be determined by trial and error using box plots and the measure of skew (Equation 3.25) to support the process. For example, if

Table 3.7 The ladder of transformation for skewed data

| Stronger | Correct negative skew | | No change | Correct positive skew | | Stronger |
	Mild change			Mild change		
Antilog(x)	x^3	x^2	x	$\sqrt{x}$	log(x) $-1/x$		$-1/x^2$

Sources: Erickson and Nosanchuk (1992), after Mosteller and Tukey (1977).

a variable, x, exhibited mild positive skew, we could try calculating $\sqrt{x}$, $\log_{10}x$ (or $\log_e x$) and $-1/x$, and then comparing the results.

For the income data, using the common logarithm ($\log_{10}x$) works well. The resulting distribution is still not perfectly normal but is more symmetrical around its centre (Figure 3.14). The skew value is now Sk = 0.288.

The long tail

A phenomenon described as 'the long tail' (Anderson 2006) has gained attention in economics, marketing and business studies. The shape of the distribution is exemplified by Zipf's rank-size rule for cities in a nation or some other economic system (Zipf

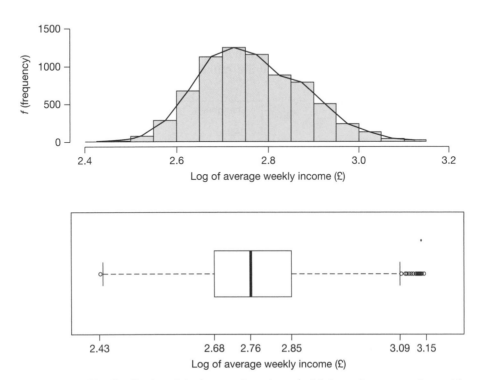

Figure 3.14 The distribution of the income data after a 'mild change' to correct the positive skew. The skew value is now Sk = 0.288

> ### Key concept 3.8   Skewed data
>
> **Skewed data** have a tail of more extreme values to the right or the left of the mean. If it is to the right then the data are said to be *positively* skewed and the mean will be greater than the median. If the tail is to the left, the data are *negatively* skewed and the mean will be less than the median.
>
> Social data sets, including census variables, are often positively skewed. This is due to the way the data are bounded and because the average tends to be nearer to the minimum possible value (0%) than to the maximum (100%).
>
> It is often helpful to **transform** skewed data using 'the ladder of transformation'.

1941; 1949). In loose terms the rule states that if the population of a town is multiplied by its rank, then this will equal the population of the largest and highest ranked city (Carter 1995, p.35). Expressed more formally,

$$P_r = \frac{P_1}{r^q} \tag{3.26}$$

where *r* is the rank of a city, *Pr* is the population count of a city of rank *r*, P_1 is the population of the largest city and *q* is a constant, generally with a value close to one (Dicken and Lloyd 1991). Because '*r* is raised to the power of *q*' the formula results in a power-law distribution, one that has a long tail, as in Figure 3.15.

The raised interest in the long tail is due to the changing nature of commerce and the growth of online retailers. Consider, for example, book or DVD sales. A more traditional business model is to maximise profit by concentrating on the high-demand, high-turnover titles that are ranked as the most popular. However, if the marginal

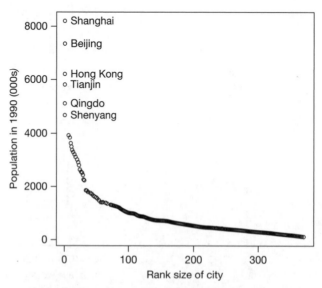

Figure 3.15   The rank-size 'rule' and 'long tail' for the population size of cities in China in 1991
Data source: http://people.few.eur.nl//vanmarrewijk/geography/zipf/, after Brakman et al. (2001).

costs of storage and distribution are much reduced for a product, as they are by digital technologies and by reaching a larger customer base, so it becomes viable – and potentially more profitable – to concentrate on the less popular titles and not the bestsellers. Consequently, a graph of book or DVD sales against their ranking will increasingly show a long tail of the lower ranked books selling in low but profitable numbers. The same arguments for niche markets can also explain the proliferation of multi-platform digital TV channels that can be watched on cell phones, computers and even (for the traditionalists) televisions.

3.10 Moments of a distribution

The measure of skew defined by Equation 3.25 is also known as the 'third moment of a sample distribution'. A clue to why, and to what the first and second moments are, is found by looking at the equation and observing that the top part of it involves a summation, whilst the bottom part is a division related to the number of observations, n:

$$Sk = \frac{\sum_{i=1}^{n}(x_i - \bar{x})^3}{(n-1)\,s^3} \tag{3.25}$$

The same pattern can be observed in how the mean and the variance are calculated:

$$\bar{x} = \frac{\sum_{i=1}^{n} x_i}{n} \qquad \text{(a summation and a division by } n\text{)} \tag{2.3}$$

$$s^2 = \frac{\sum_{i=1}^{n}(x_i - \bar{x})^2}{n-1} \qquad \text{(a summation and a division related to } n\text{)} \tag{2.7}$$

There is a fourth moment which measures **kurtosis** (Ku), which is how flat or peaked a distribution is:

$$Ku = \frac{\sum_{i=1}^{n}(x_i - \bar{x})^4}{(n-1)\,s^4} - 3 \tag{3.27}$$

The subtraction of 3 in Equation 3.27 gives a normal curve a kurtosis value of zero. By comparison, a Ku value about zero indicates a peaked distribution (described as leptokurtic) whereas a value below zero indicates a flat distribution (platykurtic) – see Figure 3.16.

Together the four moments of a distribution give a more complete description of the shape of a data set than the mean and the variance. Whilst it is the first two moments, the measures of centre and of spread, that will be the most important as we go further in this book, the measures of skew and of kurtosis are useful diagnostics for checking the data are as normal as they appear.

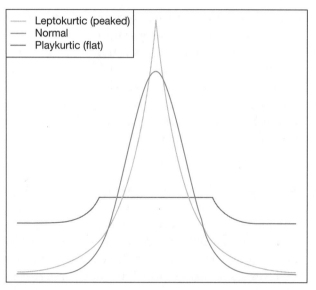

Figure 3.16   A peaked (leptokurtic), a normal and a flat (platykurtic) distribution

---

Key concept 3.9   Moments of a distribution

The four moments of a distribution can be used to summarise the shape of a data set and to check for normality.

The **first moment** is the **mean** and indicates the centre of the data.

The **second moment** is the **variance** and suggests how spread out the data are around their centre. The square root of the variance is the standard deviation.

The **third moment** is a measure of **skew** and can be used to detect a tail of relatively high or low values.

The **fourth moment** measures **kurtosis** – how flat or peaked the distribution is.

3.11 The quantile plot

Throughout this chapter, visual methods have been important tools when checking for normality. Specifically, histograms have been used to convey the overall shape of the data, and box plots to identify outliers. Both are useful, especially for initial exploration of the data (gaining familiarity with the data and checking for error). But both are flawed: the histogram because the number and the width of the bins are arbitrary; the box plot because it says little about the shape of the data within the IQR.

A better visualisation for looking at whether the data are normal is a quantile plot. The ideas behind this are based on the concept of quantiles (introduced in Chapter 2) and on the properties of a normal curve.

To explain the quantile plot, we start by taking the MPs' expenses data (here including the three outliers), ranking the data from lowest to highest, and converting the ranked data into z values. We then find out what proportion of the data has values less than each of the ranked values.

Table 3.8 The data used to produce the quantile plots. The explanation is given in the text

Rank position of the data value within the set (k)	1 (min.)	2	3	4	5	. . .	643	644	645 (max.)
x (£)	44551	67891	82286	91737	92986	. . .	173691	178116	185421
z (from the data):									
$(x-\bar{x})/s$	−5.9253	−4.4105	−3.4763	−2.3674	−2.3937	. . .	2.4559	2.7431	3.2172
$p=(k-1)/(n-1)$	0.0000	0.0016	0.0031	0.0047	0.0062	. . .	0.9969	0.9984	1.0000
z (from the standard normal)	−∞	−2.9571	−2.7364	−2.6002	−2.4999	. . .	2.7364	2.9571	+∞
$p=k/(1+n)$	0.0015	0.0031	0.0046	0.0062	0.0077	. . .	0.9954	0.9969	0.9985
z (from the standard normal)	−2.9580	−2.7374	−2.6013	−2.5010	−2.4210	. . .	2.6013	2.7374	2.9580

For example, the minimum value for the expenses data is £44551, which has a z value of −5.9253. Since this is the lowest value in the data, no others can be lower.

For the second-ranked value, where $x=67891$ and $z=-4.4105$, we now know there is one lower value. That is, one of 645, which as a proportion is $p=1/645=0.0016$.

For the third-ranked value, $z=-3.4763$ and the proportion is $p=2/645=0.0031$. Continuing in this way for all the ranked values gives the second and third rows of Table 3.8 (excluding the row of ranked values).

The key to the quantile plot is to take each of the p values calculated from the data, look at its associated z value, and compare that z value with what it would be if the data were truly normal.

The first value in Table 3.8 has $z=-5.9253$ for $p=0$. Under the standard normal curve, the position that has $p=0$ of the curve to the left of it is at $z=-\infty$.

The second value has $z=-4.4105$ for $p=0.0016$. Under the standard normal, the position that has $p=0.0016$ of the curve to the left of it is $z=-2.9571$.

The third value has $z=-3.4763$ for $p=0.0031$. The true value for a normal curve is $z=-2.7364$.

Looking down the columns of Table 3.8, if the expenses data were perfectly normal then the z values for the data would be the same as the z values for the standard normal, and all the points would lie on the straight line shown in Figure 3.17.

There is a problem with producing a quantile plot in the way described, evident in the fourth row of Table 3.8. The minimum and the maximum values of the data set cannot be shown on the plot because their p values of 0 and 1, respectively, correspond to z values of negative and positive infinity for the standard normal curve. To get around the problem, approximate values for p may be calculated so that alternate values of z can be obtained from the standard normal and shown on the plot instead. One way of obtaining the approximations is shown in the fifth row of Table 3.8.

Finally, we take the opportunity to plot the original data instead of their z values on the vertical axis of the quantile plot. Despite this, and despite the approximations, the principle remains the same. If the expenses data were normally distributed then all the points would be on (or, at least, close to) the straight line, now shown in Figure 3.18. Instead, the presence of the three outliers – the MPs with expenses less than £90000 – is evident.

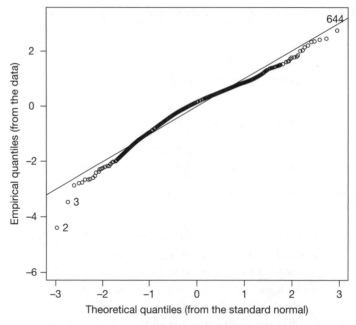

Figure 3.17 A quantile plot of the expenses data, based on the second and fourth rows of Table 3.8. This is not the best way of producing a quantile plot, however, because the lowest and highest data values (ranked 1 and 645) cannot be shown. See text for details

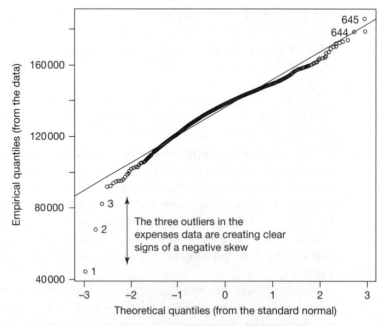

Figure 3.18 A better quantile plot of the expenses data. All the data are shown; the skewing effect of the three outliers is evident

Interpreting the quantile plot

More important than knowing how the quantile plot is constructed is to know what it looks like in the presence of outliers, skew and other deviations from normality. To this end, Figure 3.19 shows that:

(a) for negatively skewed data, the points appear to curve beneath the straight line;

(b) for positively skewed data, the points appear to curve above the straight line;

(c) for strongly peaked data, the points curve either side of the straight line; and

(d) for approximately normal data, the points lie on the straight line.

> **Key concept 3.10   The quantile plot**
>
> A **quantile plot** is used to check for signs of non-normality, such as the presence of outliers or skew.
>
> The quantile plot is better than a histogram because it does not group the data into bins of arbitrary width and number. It also offers advantages over the box plot because it shows all the data, not just the IQR, median and outliers.

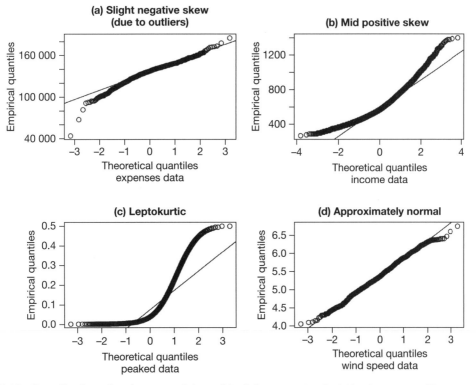

Figure 3.19 Quantile plots of various sets of data with slight or greater deviation from normality

3.12 Conclusion

This chapter has focused on the normal distribution. We applied some of the descriptive, summary and visual techniques learned in Chapter 2, to look at the shape of data and to detect skew and outliers.

We used the normal curve as a model for our data. Based on its mathematical properties we were able to assess the probability of drawing a particular range of values from the normal curve, to assess how unusual or otherwise the result would be over the long term.

In doing so, we are beginning to move beyond interest in the data for their own sake, to asking questions about the structures and processes that generate the data. In the language of sampling that is introduced in the next chapter, we are interested not just in the sample of data but also in what the properties of the sample reveal about the population from which the sample was drawn.

Key points

- A normal distribution has a humped or bell-shaped appearance, with more of the values found closer to the centre than towards either end.
- The shape of a normal curve is defined by its mean and its standard deviation. It is symmetrical around its mean.
- The central limit theorem suggests that many processes will generate normal data.
- The properties of a normal curve are that 95% of its area is within 1.96 standard deviations of the mean, and 99% is within 2.58 standard deviations.
- We can use a statistical table, combined with the 'left to right and subtract' method, to calculate other areas under a normal curve.
- We can use the properties of a normal curve to determine the probability of drawing one from a chosen range of values from a normally distributed (random) variable.
- Here, probability indicates what would happen in the long run.
- Data are standardised if their original measurement units are converted into units of standard deviation from the mean. The resulting values are called z values. The shape of the data does not change.
- Not all processes generate normally shaped data. Signs of non-normality include the presence of outliers, skew and kurtosis.
- It is often possible to use one from the ladder of transformation to return skewed data to normal.
- Quantile plots provide a useful visual method to check the normality of data.

References

Albert, J. (2007) *Bayesian Computation with R,* New York: Springer.

Anderson, C. (2006) *The Long Tail: Why the Future of Business Is Selling Less of More,* New York: Hyperion.

Avila, V.L. (1995) *Biology: Investigating Life on Earth,* London: Jones and Bartlett.

Bolstad, W.M. (2007) *Introduction to Bayesian Statistics,* 2nd edn, Hoboken, NJ: Wiley.

Brakman, S., Garretsen, H. and Marrewijk, C.V. (2001) *An Introduction to Geographical Economics: Trade, Location and Growth,* Cambridge: Cambridge University Press.

Carter, H. (1995) *The Study of Urban Geography,* 4th edn, London: Hodder Arnold.

Dicken, P. and Lloyd, P.E. (1991) *Location in Space: Theoretical Perspectives in Economic Geography,* 3rd edn, New York: Prentice Hall.

Dorling, D., Rigby, J. and Wheeler, B. (2007) *Poverty, Wealth and Place in Britain, 1968 to 2005,* Bristol: Policy Press.

Erickson, B.H. and Nosanchuk, T.A. (1992) *Understanding Data,* 2nd edn, Maidenhead: Open University Press.

Frank, H. and Althoen, S.C. (1994) *Statistics: Concepts and Applications Workbook,* Cambridge: Cambridge University Press.

Galton, F. (1886) Regression towards mediocrity in hereditary stature. *Journal of the Anthropological Institute,* 15, 246–263.

Glass, L. and Tobler, W.R. (1971) Uniform distribution of objects in a homogeneous field: cities on a plain. *Nature,* 233(5314), 67–68.

Grinstead, C.M. and Snell, J.L. (2006) *Introduction to Probability,* Providence, RI: American Mathematical Society. Available at: http://math.dartmouth.edu/~prob/prob/prob.pdf, accessed 20 October 2010.

Kreft, I.G.G. and Leeuw, J.D. (1998) *Introducing Multilevel Modeling,* London: Sage.

Kruskal, W. (1978) Formulas, numbers, words: statistics in prose. *American Scholar,* 47(2), 223–229.

Mesev, V. (2003) *Remotely-Sensed Cities,* Boca Raton, FL: CRC Press.

Mosteller, F. and Tukey, J.W. (1977) *Data Analysis and Regression: A Second Course in Statistics,* Reading, MA: Addison-Wesley.

Pearson, K. (1924) Historical note on the origin of the normal curve of errors. *Biometrika,* 16(3/4), 402–404.

Silventoinen, K. (2003) Determinants of variation in adult body height. *Journal of Biosocial Science,* 35(02), 263–285.

Zipf, G.K. (1941) *National Unity and Disunity: The nation as a bio-social organism,* Bloomington, IL: Principia Press.

Zipf, G.K. (1949) *Human Behavior and the Principle of Least Effort: An Introduction to Human Ecology,* Cambridge, MA: Addison-Wesley.

Sampling

Chapter overview

In geographical research, as with other subject areas, it is not practical to collect all data relating to a task for logistical reasons, for example budgets, time and access. Fortunately, neither is it necessary in order to draw understandings about a process or phenomenon, given an appropriate sample or subset of data. In this chapter, we outline what we mean by an 'appropriate' sample, and discuss how you should go about collecting a sample that is fit for your particular research task.

A theme within this chapter is that a clear research question and focused scope for a study are vital prior to considering the sample design. A good study will be backed up by reading that establishes the factors that might potentially affect the answer to your research question. It is not feasible to consider all such factors within one research project; rather, you need to design the way in which you collect your data sample such that you are able to establish an answer to one clear question relating to a defined area and varying factor. We outline a number of probabilistic and non-probabilistic methods for selecting a sample that are commonly used in geographical studies. To be able to draw more general conclusions from your study (Chapter 5), probabilistic sampling methods are needed to ensure that the sample you collect has similar characteristics to that from which you draw your measurements.

Additionally, the chapter highlights that sample design needs to be considered both from statistical and practical perspectives; issues relating to the data processing chain and measurement accuracy can impact on the subsequent usability of the data.

Learning objectives

By the end of this chapter, you will be able to:

- Know the distinction between a sample and the population.
- Explain and recognise examples of sampling bias and the notion of a representative sample.
- Outline the stages involved in the process of sample design.
- Distinguish between non-probabilistic and probabilistic sampling methods.
- Summarise the nature of the following non-probabilistic sampling methods – judgemental, quota, snowball and convenience sampling.
- Explain the concept of a sampling frame and be able to apply this to a study of your design.
- Summarise the following probabilistic sampling methods – systematic, simple random, stratified random, multi-stage random and cluster sampling.
- Highlight the effectiveness of different sampling methods and be able to construct a sample design that is effective and efficient for a given geographical problem.
- Appreciate the relationship between sample size, precision and confidence in general terms.
- Weigh up the practicalities of a particular sample design, taking into account cost, safety, access and environmental/human ethics.

4.1 Introduction

In geographical research, as with other subject areas, it is often simply impractical to collect all data relating to a task for logistical reasons including budgets, time and access. We cannot, for example, interview every adult within an electoral district, measure the girth of every tree within a forest or count every grain of sand on a beach. Moreover, as you will see by the end of this chapter, this would not be a very efficient way of going about your task even if a full survey (or census) was feasible. Largely, we do not need to gather a complete and full set of data relating to our task in order to draw general conclusions about the processes or phenomena we are investigating. There are always exceptions to any generalisation, and there are certain circumstances where it is possible and feasible to deal with a complete population, although these situations are rare. You could for example interview every individual on a planning committee regarding a particular decision. More normally in geographical research we may gather a sample (or subset) of data from our target population, from which we may draw robust conclusions or develop geographical theories.

> ### Key concept 4.1 Target population
>
> The **target population** can be defined as the complete set of measurements that might hypothetically be recorded in a particular study context that are relevant to the study. In a geographical context this could include for example all people or vegetation within a defined area at a certain time, or the distributed activities of a single business unit operating from one location. Often, the notion of population is more theoretical than quantifiable.
>
> A population falling outside either the geographical area or context of a study is sometimes referred to as an **out-of-scope population**.

The characteristics of a sample of a population are known as **statistics**. The nature of the population and its mathematical relationship to the sample will be discussed further in Chapter 5. In Chapter 5, too, specific mathematical notation concerning the sample and population will be introduced.

A sample that matches (or represents) the population is known as a **representative sample**. Achieving a representative sample is a key issue when adopting probabilistic statistical methods in your research, but also applies when adopting other geographical methodologies whether qualitative or quantitative in approach. However, not all research traditions require that the outcomes are representative of the whole, as some work is illustrative in nature. It is important to respect these differences across the discipline; there is a place for the in-depth case study, particularly in regard to gaining more detailed knowledge about how a process or phenomenon emerges. This caveat applies in both human and physical geography.

> ### Key concept 4.2 Representative sample
>
> This is a sample that matches (or represents) the statistical characteristics of the overall target population. The characteristics of a sample are known as its **statistics**.

The stages of the sampling process are outlined within Figure 4.1. This chapter begins by looking in more depth at the general steps that need to be taken before constructing a sampling plan. A considered and focused research question is a key start to this process (Section 4.3), as is a review of the related literature to find out what factors might influence the variable you are sampling. Together, these points feed into a review of the scope of your study to make sure that the work looks manageable with the time and resources you have available. With these tasks in hand, the next step in the sampling process is to define your target population, an operation known as constructing your sampling frame. Once these issues have been considered, the first stage of the sampling process, involving the scope and scale of the project, is complete.

In the second half of the chapter, we turn to the sampling method itself. There are many well-documented general sample design strategies, where **sample design** is the method by which you select a subset or sample of data for a particular purpose.

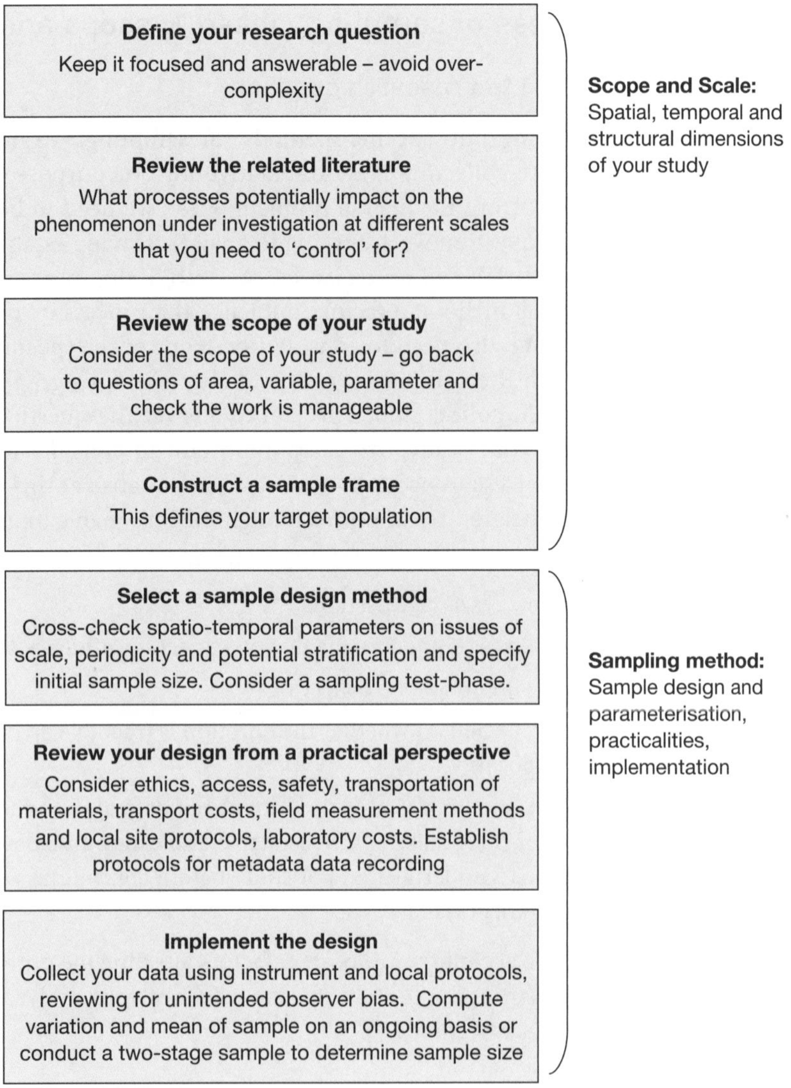

Figure 4.1   The process of sampling

Considering which design best meets the goals of your study forms an important next step in the process, with the underlying aim behind the process of sample design being to ensure that the subset of data collected reflects the characteristics of the overall target **population**. Sub-questions here include how the design should be parameterised, and whether you should first conduct a pilot study. Theoretical ideals and practical realities conflict; the chapter therefore also includes some practical pointers that may suggest that your sampling strategy needs modification before you go out in the field. The chapter is of clear relevance to those who are going to undertake primary data collection and analysis. In addition, principles of sampling are important for users of secondary data to understand too. Any analysis is dependent on data and whether they are fit for purpose.

4.2 The process of sampling, phase 1: scope and scale

Formulating the research question

Before jumping into specific methods for sampling, we need to be sure that the sample we are collecting is representative for our purpose. At this point, we need to stand back from the overall problem. First, we need to be absolutely clear about the research question we are trying to answer. Having established this fundamental step in the overall, and much larger, research design process, we need to ask which environmental or human factors influence the process or phenomenon we wish to investigate. We also need to consider the scope of the specific study; is it feasible, in terms of both the complexity of the question and geographical coverage intended that you could collect sufficient data to answer the question in the time you have available for your study? It is also important to consider at what scale you might expect the process to occur or pattern to materialise at this early stage in the study and what associated assumptions it might be reasonable to make regarding the data. To recap:

- What is your research question?
- What processes or phenomena do you expect to influence that question?
- What data should you be collecting?
- Have you reflected on whether the question is tractable given the time and resources at your disposal?
- How much do you already know about the data and underlying geographical process? Rather than making unreasonable and ungrounded assumptions, it would be appropriate to undertake a pilot study before continuing with your main sampling campaign if necessary.

Omitting to account for important factors affecting the problem you wish to address can have disastrous implications for the results of your study, as a result of introducing **sample bias** (see Section 4.3).

Review the relevant literature

The range of potential issues that geographers and environmental scientists might need to consider when reflecting on factors affecting a spatial process or phenomenon is as broad as geography itself. There is no substitute for detailed background literature work prior to developing a research methodology and statistical sampling design. Commonly encountered factors, subject to the study in question, might include: gender, sexual preference, family background, cultural background, economic background, educational background, computer literacy and access, age, niche personal interests, political preferences, subsurface or deep geology, longer-term environmental history of an area (for example, flooding, fire, uplift), meteorology and/or climate, depth, elevation, soil type, land cover, land use, basin size, distance from the sea, aspect, slope.

This long, but hardly comprehensive, list serves to illustrate the need to constrain the scope of your study. The tighter your reference question, the easier it is to avoid a biased sample.

The scope of your study

Formulating precisely a research question that is as specific as possible is the easiest way of managing the scope of your study. The scope of your study can be further tightened in two main ways. First, and most simply, by reducing the geographical extent of your study, and secondly by choosing sample sites such that a number of potentially important factors affecting the process you are observing are held steady. This second practice is known as controlling your variables.

Geographical extent

A look at the titles or abstracts of journal papers in areas of the discipline that particularly interest you can be instructive here. Taking one example of how you might construct a focused dissertation title, consider the following title: 'The geography of shoplifting in a British city: evidence from Cardiff' (Bromley and Thomas 1999). The title shows that the authors are interested in the phenomenon of British shoplifting as a whole, but indicates with clarity that their evidence in regard to the complete country is partial and geographically specific.

Controlling variables

A solid piece of research looking at the effect of geology on river incision is better managed by comparing two basins of similar size and rainfall patterns than trying to undertake a complex model that accounts for varying basin size, rainfall volume and rainfall intensity in one go. This is because achieving a representative sample in the latter case would be a sizeable task before you could achieve any statistical certainty in your results, owing to the multiple potential interactions between variables. Similarly, you might choose to investigate the impact of an environmental variable (for example, distance from a power station or mobile phone mast) on health by comparing groups of similar age, economic and genetic background who have lived in the area for similar lengths of time.

In many aspects touching on geography, from meteorological recording to water quality, local site protocols for sampling are in regular use and you should conduct a careful investigation of your subject domain. Where mid-scale processes are your target, these local variations should be controlled carefully also. The factors held steady as part of the statistical experiment are known as controlling factors. The need to untangle complex mechanisms can be avoided by streamlining your research design and keeping it simple.

Questions of scale

The question of scale is an overriding theme throughout geography (see Chapter 8). The most important thing to consider from a sampling perspective is the scale at which you anticipate the phenomenon you are investigating to vary the most. In many cases, variation should be considered over both space *and* time. Let us take the biodiversity of different components of a shoreline as an example. One might wish to record data during the periods of highest expected diversity in the season; further, owing to the nature of the biological specimens, one would need to record rocky areas populated by microalgae in more spatial detail than areas of soft sediment populated by seagrass. Do not forget also that the scale of a phenomenon may be variable with direction; drumlins, for example, are elongated features and a drumlin field tends to form along a particular axis. A paper by Smith *et al.* (2006) on the geomorphological mapping of glacial landforms highlights both this issue of symmetry and directionality and considers the effects of using secondary data at different scales on the mapped features.

Sampling frame

The **sampling frame** contains all possible data (or **sampling units**) to be selected (the population); it frames or outlines the data set. Only data within the sampling frame may subsequently be selected for analysis. The term sampling frame is more commonly found in human than physical geography, but the principle applies to all practices of sampling.

Key concept 4.3 Sampling frame

This is the actual list from which your sample items are drawn. It might be the GIS data grid subdividing your study area in the context of a physical geography example, or the rather more tangible electoral roll (register of voters) in the case of a human geography task.

Key concept 4.4 Hierarchy of terms in probability sampling

We have outlined the population, target population, sample frame and sample. These relate as follows:

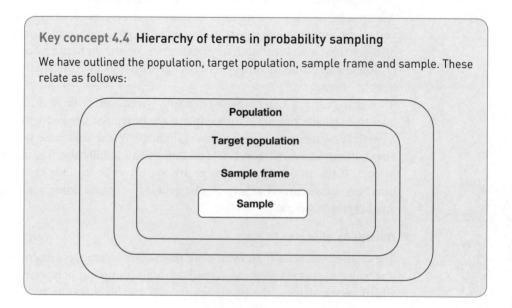

Population

Target population

Sample frame

Sample

Examples of sampling frames include the electoral roll for an area, the list of students matriculated at a university or school, or the complete police record of all shoplifting crime reported in a particular place or timespan. Sampling frames may also be geographical; consider the use of the electoral ward or enumeration district, a list of postcodes in an area or another form of regional boundary. In the case of physical geography, the sampling frame could be the list of operating meteorological stations or pollution control instruments in an area, a watershed relating to a river feature under investigation or a digital map showing geological or elevation classifications for the study area under review. Sample units should be defined by size (for example, quadrat extent in a bio-geographical study) and the location and time of records should also be noted.

4.3 The process of sampling, phase 1: issues

Efficiency issues

Collecting, and analysing, data are both time-consuming processes. Whilst human nature appears to encourage a tendency to collect data 'because they might be useful', without careful planning it is quite easy to amount volumes of unusable data and miss out something that later turns out to be critical. Bearing in mind the issues raised in Section 4.2, consider your answers to the questions in the box below.

Review scenarios and questions: Scope, scale and extent

- If you are looking at biodiversity indicators, do you need to record the individual proportions of named species in your sample notebook? Might it be better to cover a wider geographical range, noting number of different species, instead?

- You are seeking data about shoplifting in the UK. Should you rely on secondary data concerning recorded shoplifting incidents, given that these are suggested to be but a small sample of the whole (Bromley and Thomas 1999), or should you weigh up other strategies to gather your data?

- You are interested in the impact of green lanes on butterfly distributions. Butterflies are in general most active at high levels of sunshine. Is it really worth undertaking a sampling campaign at both 9am and 12pm, or would one set of observations at 12pm be just as effective?

- You have been investigating the effects of heavy industry on the quality of river water. While you are there, you decide it might be interesting to look at levels of phosphorus and nitrogen also. After you have analysed the different chemicals, at considerable time and expense, you realise that you have no data concerning the history of agricultural land use in the area and cannot make good use of your nitrogen and phosphorus data. Further, because you have undertaken more analysis than planned, your project may run late and runs the risk of not being completed. Think back, what might you have done differently?

In general, these scenarios arise for a variety of reasons. First, there is doubt regarding the research question itself or a lack of consideration concerning the processes and variables influencing your research question. Secondly, focus and discipline are required as part of an effective sampling campaign in the fieldwork. This second issue is more likely to trap the enthusiastic and diligent researcher, determined to succeed. There is also a danger in collecting data because they are what has been collected in the past, affordable, familiar or easy to collect.

The problem of sampling bias

We have already established that a sample collected for the purpose of drawing statistical inference should reflect the salient characteristics of the overall population, for example age, gender, economic or cultural background for human participants, or elevation, geology or climate in the case of many physical geography projects (Key concept 4.5). The word salient is key here; the characteristics that matter will vary according to the purpose for which the sample is being collected. As part of sample design, you must consider which factors are likely to affect the particular process or phenomenon that you are measuring and avoid over-representing one group of environmental conditions or people over another.

Why does this issue of bias matter?

Bias in the *Oxford English Dictionary* is defined as follows: 'An opinion, feeling or influence that strongly favours one side in an argument or one item in a group or series.' In a scientific sense, your intuition or opinion may be correct, but it is vital that you do not invalidate your research by rigging the data in a way that favours the outcome you desire, intentionally or otherwise.

Key concept 4.5 Sampling bias

Sampling bias (also termed **situational bias** and **situation specifity**) occurs when the sample is, in fact, unrepresentative of the target population; it favours some elements over others. In geographical terms participants should reflect the salient characteristics of the population about which inference is being drawn. Sampling bias most often occurs through a lack of forethought rather than the deliberate manipulation of data, and can arise for a variety of reasons:

- Bias in regard to geographical site or situation and the overall size and controlling factors behind your research question; these could relate to water supply, elevation or geology in a physical geography setting or cultural and economic factors in a human geography context.

- Bias introduced when sampling at an unrepresentative time/or period.

- Bias introduced through the method you use to collect your data.

- Too small a data set, even when important controlling factors have been weighed up as part of the sample design, can lead to bias where the underlying variability of the population is relatively high compared with the sample size.

At the root of the phrase 'Lies, damned lies, and statistics' (attributed to Benjamin Disraeli) is this issue of bias; Disraeli's argument may have related to the manipulation of state figures rather than scientific statistics, but the underlying issue is identical. If the sample is biased the results that you might subsequently claim would lack credibility, be unpublishable and have a high chance of being misleading.

We will consider how best to approach the question of sample design, with a view to showing you how to gather representative and non-biased data in Section 4.2. Before doing so, let us look at some examples of bias to illustrate this issue.

Sample bias: Example 1

> #### Exemplar - sampling bias: Scenario 1
>
> An article was written for a newspaper in the late 1980s concerning the drinking habits of students. This story did actually run, but is reported here anecdotally and from memory. The gist of the piece was that, on the basis of evidence from approximately 40 students from one particular college of one particular university, all students spent a high proportion of their government grant on gin and tonic; hence, the government grant to students should be abolished. Prima facie this not a geographical topic, but if we look carefully there are a number of space–time issues that emerge that are common to many more overtly geographical research projects.

If we unpack the story outlined in the exemplar box above from a common-sense perspective, intuitively we will realise the following flaws in the sampling plan that led to an unlikely conclusion:

- We do not know whether the proportion of males to females in this study matched that of the overall student population in the UK at that time, but given that in the 1980s female drinking was less prevalent than it is now, this proportion could have been material even to the outcome of the small study.

- We do not know the religious background of these students. The consumption of alcohol is not a predominant feature in Muslim society, for example, so in this particular case the religion of the students is a salient issue.

- Both the gin and tonic association and the particular name of the college and indeed university at the time in question were, in a British context, suggestive of a particularly privileged economic background amongst this student group. To suggest that the results were indicative of the behaviour of all students was to extrapolate the findings beyond the supporting evidence.

- Not all students drink gin and tonic as a matter of preference, wherever they study. Were the data collected in a bar particularly known for its range of spirits? In other words, were the data collected at a representative range of sites?

- Were the data collected in a bar at a time when gin and tonic was on special offer, or in a week of celebratory significance? In other words, were the data temporally representative of the 'norm'?

- In a British context, and as a generalisation, gin and tonic is a combination of drinks more commonly (if not necessarily) served in the south of the country. The volume of all alcoholic drinks consumed, or even all spirits, would have been more representative of the nation's student population as a whole.

- Forty students was a desperately small proportion of the national student community at that time; the volume of data in comparison with the overall scene you wish to represent matters. Sample size is not everything, but low sample sizes can certainly contribute to bias where the overall population is variable as in this case.

- The article itself had a tendency to support stories that favour a right-of-centre political viewpoint. That is, the results appeared to have been used to support 'An opinion, feeling or influence that strongly favours one side in an argument.'

This all adds up to a case of **biased** statistics: too few students who in all likelihood were from similar economic and religious backgrounds, of unknown gender, compounded by geographical bias at the micro scale (one bar) and macro scale (southern England being used to represent the UK). We will look at how much data you should collect later in this chapter (Section 4.5.4); estimating approximately what size of sample you might need when planning a data collection campaign, and working out when to stop sampling, are important components within the data gathering section of a project.

Sample bias: Example 2

> #### Exemplar – Sampling bias: Scenario 2
>
> In 1948, the *Chicago Daily Tribune* was so confident of its opinion polls that it went to print with the title 'Dewey beats Truman'; but Truman won. The public opinion polls prior to the British 1992 election are a more recent example of statistical disaster as a result of bias (Smith 1996).

Even when great care is taken designing samples to avoid causes of bias, the unexpected can happen as shown in the second example, above. The cause of the highly embarrassing mistake in the *Tribune* case was that the opinion polls were based on telephone surveys, but in 1948 many people with lower incomes did not own telephones (McAfee 2002, p.226). George W. Bush too was predicted to lose the 2005 US election with a heavy defeat according to the exit polls. In fact, he won convincingly.

Bringing this scenario into the present, if you are undertaking a geographical project today, for example looking at the effectiveness of web media for communicating local issues, or evaluating the effectiveness of public participatory GIS (PPGIS) using visual media for community decision making, you should consider whether you need to seek out those without easy access to the Internet as well as those who can connect easily. Broadband access is not uniform geographically, economically or across age-cohorts universal.

In both the exemplars above, the data were collected for the direct purpose of the studies reported. Such data are referred to as **primary data**. Data gathered by a researcher as part of a study, whether by interview survey, experiment or observation, fall into this category. In contrast, **secondary data** are data that have been collected as part of a separate study and potentially a different purpose, but that offer potential value to your work. These data might for example include published government statistics, meteorological records, elevation surfaces and remotely sensed land cover classifications (Chapter 2, Section 2.8). Because you have not controlled the collection of these data in order to make sure that the sample is representative in regard to the process that you yourself are researching, the probability of sample bias creeping into a study using secondary data is strong. Whether the data set is **fit for use** (Key concept 4.7) in your particular study must be carefully weighed.

Sample bias: Example 3

Let us consider the UK Meteorological Office land surface temperature archive as an example of a secondary data set that we wish to consider for the purpose of creating raster maps at 1 km resolution for climate variables over England and Wales. We also wish to compute associated summary descriptive statistics for England and Wales as part of our project.

Key concept 4.6 Metadata

Metadata are formally described as 'data about data'. The term encompasses information such as the date of data collection, the age or address of the informant, spatial resolution, attribute precision, sample size, method by which data were acquired through processing in the lab and the type of instrument used to make the recording.

Like other national meteorological networks, great care is taken by the UK Meteorological Office that measurements adopted within the archive use instruments calibrated to a certain accuracy, based on equipment at sites conforming to carefully specified local characteristics and using well-documented observing and recording protocols (Met Office 2010). A very positive aspect to this particular secondary data collection is that the **metadata** associated with it are strong; we know what we have, and how and when it was recorded. This is to ensure that users can evaluate whether the data, and their consistent protocols, are fit for the purpose they need them for.

Key concept 4.7 Fitness for purpose/use

This is the effectiveness and appropriateness of the data *for the particular task for which they are to be used*. Factors affecting fitness include where and when the data were collected and by whom (expert or amateur; scale and place of collection; numerical precision of record). Key here is the interplay between defined purpose and characteristics of the data.

Let us look as an example at the locations for which daily temperature data are available for a period we wish to study against the national picture for England and Wales (Jarvis 2000). The mean elevation of 174 sample meteorological recording stations, scattered across England and Wales, was 83 m. Based on a GIS analysis using Ordnance Survey 50 m raster elevation data, and using the central point of each 1 km cell across the landscape of England and Wales, the mean elevation for England and Wales was computed at 125 m. Further, the standard deviation in elevation within a 50 km^2 area around each 1 km cell across the country was 47m based on the sample points and 60m for the overall landscape at 1 km resolution. These figures show that the Met Office collection is biased towards lower and flatter elevations relative to the overall situation for England and Wales. An average temperature for the total area taken from the subset of 174 archival records, is likely to be higher than reality. Why might this be so?

The primary purpose behind these national meteorological records is the forecasting of daily weather. Accurate weather forecasts are particularly pertinent for the building and agricultural industries and for air travel. Thus, many meteorological stations are based at airports or airfields, generally not sited in mountainous regions; local sources of meteorological data are more commonly available in lowland agricultural regions, again which tend to be at lower elevations. We might conclude that the data archive is fit for its main purpose, but is it fit for our purpose (Key concept 4.7)?

The answer to this question depends to a large degree on the intended purpose for the temperature surfaces themselves; the data *are* biased, but if we determine that the temperature surfaces are to be used for agriculture-related modelling, the outputs for which are less relevant in mountainous areas, then we can still conclude that the data set is sufficiently representative for our task. The geographical coverage, longitudinal nature of the data set (data collected over a number of time periods, in this example multiple years) and the quality assurance applied to the archive also contribute to this assessment.

Increasingly, data collected as part of large collaborative scientific experiments are archived for access from the Internet, yet, as you may discover for yourself, supporting metadata are often sparse. This makes bias more difficult to spot. Overall, however, the same questions you need to consider in regard to sample design for primary data should be asked of a secondary data source. If critical questions cannot be answered directly or by analysis as above, the data should not be used in your study. By the end of this chapter, you will understand better how to address the issue of bias up front so that your sample is fit for the purpose you intend and sampling bias is kept to a minimum.

 ## 4.4 The process of sampling, phase 2: sampling method

Once the scope, extent and controlling factors affecting a study have been examined, the next phase in the sampling process is that of sample design.

> ### Key concept 4.8 Sample design
>
> **Sample design** refers to the choice of method and the scale and detail of how you plan to implement your approach.
>
> Absolutely key to this process is having a clear research question. Before this phase, you also need to have developed a broad understanding of the factors that are likely to affect the human or physical process you are investigating through a thorough search of the literature, the geographical scope of your study, and weigh up the scale(s) at which you consider the processes are operating.

Sample design: overview of methods

Geographical sampling is time consuming and often costly. Further, the analysis of environmental samples in particular can be very expensive and also takes time. With these points in mind, sampling plans should be designed to be as effective as possible while providing as much information as realistically achievable.

Sampling methods can be divided into two camps: probabilistic and non-probabilistic methods. Taking the simpler non-probabilistic group of methods first, examples include judgemental, snowball and quota sampling (Figure 4.2, left). Where you wish to deepen your study to go beyond the descriptive statistical techniques discussed in Chapter 2 and use one of the analytical, inferential statistical techniques we discuss in Chapter 5, you will need to adopt a probabilistic sampling method in your work that allows sampling error and representativeness to be modelled. This group of methods includes systematic, simple random, stratified random, multi-stage random and clustered random methods (Figure 4.2, right).

Non-probabilistic sampling

For completeness, we will dip quickly into the nature of non-probabilistic sampling techniques. It is not advisable to use these methods when you wish to conduct inferential statistical analyses (Chapter 5). This is because models (such as probability theory) cannot be used to determine sampling error when the researcher has selected the sampling elements, making it difficult to quantify how representative the sample might be. However, where your purpose is exploratory and the statistics are intended simply to be descriptive, for difficult or rare populations, or where the development of rapport with participants is vital to your study, this class of sampling can be valuable. It is commonly of greater relevance in qualitative research, where data can only be represented on nominal scales and cannot be quantified in numeric terms (Chapter 1).

Judgemental sampling

Judgemental sampling, as it sounds, involves the researcher making sampling selections based on their prior experience to judge which samples to leave in or out to construct a representative sample overall. Where variety as opposed to representativeness is sought,

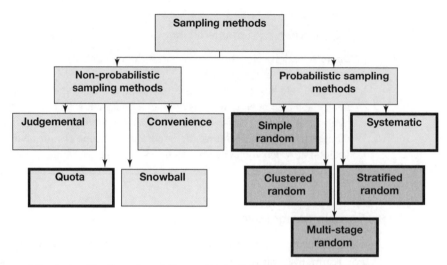

Figure 4.2 A classification of sampling methods. Dark surrounding box indicates a representative method, and grey shaded box a random method

this type of purposive sampling methodology is useful. If you were to record the opinions expressed in a public planning meeting of some type as part of your qualitative research evidence, this would gather the extent of views but with an emphasis on those most vocally expressed. One might equally make contact with a prominent local councillor to speak about the issues of a small town on the basis that that person encounters the views of many residents as part of their role.

The following excerpt from Gerald Durrell's book (1954, p.11) *Three Singles to Adventure* provides a rather amusing example of judgemental sampling:

> In a tiny bar in the back streets of Georgetown four of us sat around a table, sipping rum and ginger beer and pondering a problem. Spread on the table in front of us was a large map of Guiana, and occasionally one of us would lean forward and peer at it, frowning fiercely. Our problem was to choose a place, out of all the fascinating names on the map, to serve as a base for our first animal-collecting trip to the interior. For two hours we had been trying to make up our minds, and we still had not found a solution. I stared at the map, tracing the course of the rivers and mountains, gloating over such wonderful names as Pomeroon, Mazaruni, Kanuku, Berbice and Essequibo.
>
> 'What about New Amsterdam?' asked Smith, choosing the one really commonplace name on the map.
>
> I shuddered, Bob shook his head, and Ivan looked blank [. . .]
>
> 'What do you think, Ivan?' I asked him. 'After all, you were born here, so you ought to know the best place to get specimens.' [. . .]
>
> 'Well, sir, he began, [. . .], 'I think you'd do well if you went to Adventure'.
>
> 'Where?' asked Bob and I in unison.
>
> 'Adventure, sir,' he stabbed at the map; 'it's a small village just here, near the mouth of the Essequibo'.
>
> I looked at Smith.
>
> 'We're going to Adventure,' I said firmly. 'I must go to a place with a name like that.'

While the concept of overseas animal collection trips does not fit well with the international research ethics of today, the subsequent chapters of the autobiographical novel show clearly that the use of local expert opinion regarding the selection of a place with high biodiversity and good local cooperation worked well. In other words, where a key informant has sufficient local background knowledge about the overall population being sampled (species of Guyanese wildlife) and is articulate, and experience has shown the method to work in that case, this method can in fact be an effective way of achieving a minimally biased sample. It is, however, a method more favoured in market research than geography. The choice of sample location on the basis of an attractive place name is not advised!

Quota sampling

Quota sampling is more likely than judgemental sampling to be used as a geographical approach. As it sounds, data are added to a sample until a predefined quota is reached. If you were making a study of unemployment in relation to use or access to recreational facilities, two groups might be selected where the overall proportions of employed to unemployed between the two groups matches that of the government statistics for the area. Critically, there is no randomised element within this selection. One might target the first 200 people to arrive at the nearest railway station for a train between 8 and 8.30 in the morning, who meet important criteria relevant to the study (for example, travelling to work in London for a study on city commuting) or the first 30 people to arrive at a government job-seeking unit on a particular morning.

This method is the most likely of the non-probabilistic group to result in a representative sample being gathered. For this reason, of the non-probabilistic methods, quota sampling has most potential for use in geographical studies.

Snowball sampling

Snowball sampling involves sample members identifying other suitable candidates with particular desired characteristics, a 'chain letter' type of approach. Geographically speaking, and perhaps counter-intuitively, the effect may be to achieve a highly dispersed sample group in a geographical sense but it is very unlikely to be representative. Note that sending a questionnaire into work with a friend or parent is in reality a form of snowball sampling and such data should not subsequently form the basis for inferential tests. For populations about whom data is difficult to gather and where rapport is important, for example when considering homeless populations or activist groups, this selection-by-recommendation approach is likely to be highly valuable when used as a basis for further qualitative analysis. A quota approach could also be used to select a sample from those initially recommended by the snowball method.

Convenience sampling

Convenience sampling again operates much as it sounds. Here, accessibility is the key to selection within a trial. An academic developing a new data visualisation technique designed to improve learning might for example invite their MSc class to take part in trials regarding its effectiveness for certain tasks on a voluntary basis. The outcomes would provide initial information and feedback to the researcher, but may prove to be unrepresentative of the overall student cohort of MSc students in the subject nationally, since the convenience approach in this particular case would have a tendency to interest students particularly attracted to a visual learning style.

Our experience of students' attitude to statistics, discussed in Chapter 1, is based on a convenience sample.

Probabilistic sampling

The group of methods that follow are based on probability theory. They are key to most quantitative geographical research, and should be adopted whenever you plan to use probabilistic or inferential statistics on the data and therefore the representativeness of your data is imperative. In any of these methods, the chance of achieving an unrepresentative sample relative to the non-probabilistic methods that we have just discussed is considerably lower. Additionally, it is also possible to calculate the chance of achieving such an unrepresentative sample. We will return to this second issue when we address how many samples you need to collect to achieve your research goal (Section 4.5).

When considering geographical data in connection with probabilistic methods, it is important to note that spatial data are often autocorrelated: that is, nearby data are often more similar than those further away. Mathematically, this means that standard methods for computing sample and population means and minimum sample size (amongst others) should be adjusted. In practice, such statistical adjustments go beyond what is expected in an undergraduate curriculum. We will, however, return to the general concept of autocorrelation in Chapters 8 and 9.

Key concept 4.9 **Probabilistic methods and autocorrelated data**

Strictly speaking, **autocorrelation** in data, whether temporal or spatial, should be taken into account when computing sample statistics and drawing inference from spatial data.

Turning now to the methods themselves, the examples for the major probabilistic methods of sampling will be illustrated using a sampling frame that is a small (60 × 60 m^2) area of the Karoo in South Africa and a second larger plot in New Mexico (Dickie 2007). These data were collected as part of PhD research looking at the effect of changing vegetation on the spatial characteristics of soil in a dryland environment, comparing two contrasting sites from different continents.

The items sampled are standardised bulk density (g m^{-3}), standardised soil moisture (%) and standardised sodium levels (ppm) (Figure 4.3). The **standardisation** of data was introduced mathematically in Chapter 3 (Key concept 3.6); for now, focus on the pattern emerging from the sample and the way in which the approximation of the overall data mean changes with sample size.

Systematic sampling

Systematic sampling is a practical, simple form of sampling to get to grips with. The sampling frame is sampled regularly (systematically), starting from a randomly selected first point or element. You might for example lay out a quadrat to record bio-

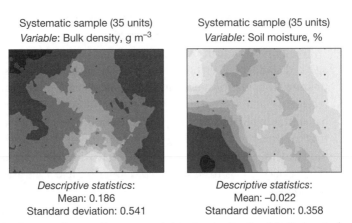

Variable: Bulk density, g m^{-3} *Variable*: Soil moisture, %   *Variable*: Sodium, ppm

Descriptive statistics:
Mean: 0.132
Standard deviation: 0.574

Descriptive statistics:
Mean: 0.051
Standard deviation: 0.362

Descriptive statistics:
Mean: 0.025
Standard deviation: 0.355

Figure 4.3   Spatial variation of standardised bulk density, soil moisture and sodium over 60 × 60 m^2 plots in the Karoo, South Africa

geographical data in a grid pattern every 5 m across your area, or select every 100th person in the electoral roll in sequence to receive your questionnaire. Your sampling frequency will be determined by the number of samples you wish to collect from the sample frame itself. Typically, the total sample frame is divided by the sample size to determine the sample interval. The sampling interval could be expressed in distance terms, or by the number of intervening records in a list.

Figure 4.4 illustrates an example of a small systematic sample being used for bulk density and soil moisture of a 60 × 60 m^2 patch of the Karoo National Park in South Africa. The sample locations are regularly placed, and in this case an identical sampling structure has been used for both variables. When compared with the background data we can see visually that coverage appears to have been achieved over all areas of significant variation.

An experiment investigating how the mean and standard deviation of our sample varies with sample size can be seen in Figures 4.5 and 4.6 respectively. Note that the

Systematic sample (35 units)
Variable: Bulk density, g m^{-3}

Systematic sample (35 units)
Variable: Soil moisture, %

Descriptive statistics:
Mean: 0.186
Standard deviation: 0.541

Descriptive statistics:
Mean: −0.022
Standard deviation: 0.358

Figure 4.4   Systematic sampling approach for standardised soil bulk density (g m^{-3}) and soil moisture (%) across a 60 × 60 m^2 sampling frame in the Karoo National Park, South Africa

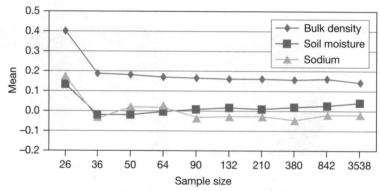

Figure 4.5 Variation of sample average created using a systematic methodology, by sample size, for variables bulk density, soil moisture and sodium

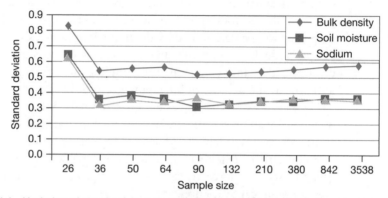

Figure 4.6 Variation of standard deviation within the sample created using a systematic methodology, by sample size, for variables bulk density, soil moisture and sodium

axes of these graphs are non-linear; they have been stretched to emphasise the flattening of the variation with relatively small sample sizes.

The graphs confirm our intuitive impressions; at a sample size of 36, both average and standard deviations are startling to stabilise and approach the values of the unknown population mean and standard deviation. The sample would be more representative at a sample size of 90, but beyond this level there would be little to gain by continuing to add data to your collection. The plot for standard deviation suggests that continuation of sampling beyond the initial 36 units would be most beneficial for bulk density. Given that this is the variable with most variability within its population (Figure 4.4), this is to be expected. Larger sample sizes are needed for more variable phenomena.

The major difficulty when using systematic sampling arises in relation to periodicity within the underlying phenomenon. For example, it is a common cultural phenomenon in the UK for houses on one side of a street to receive odd numbers and the other, even numbers. An extreme outcome might be that if you sampled every other household in a street systematically, you might find that your sample covers every house on one side of the street at the expense of the other. In a physical geography setting, terrain might vary systematically also; consider the drumlin fields in western Scotland mapped by Smith *et al.* (2006) in which terrain undulates approximately every 300 m in a classic 'basket of eggs' topography. In other words, systematic sampling will cause bias where the underlying structure of the phenomenon being measured has the same spatial structure as that of the sample interval, or indeed the phenomenon measured varies at a smaller interval than that chosen for measurement.

Simple random

Using this simple method, and assuming no autocorrelation within the data (Key concept 4.9), there is an equal chance of selecting one element of the population from another. In practice, at undergraduate level we will ignore the statistical caveat regarding autocorrelation. Selection is made via a random number algorithm, and without replacement.

There are a number of software packages that can help with the selection of a random pattern over space. These include the remote sensing packages Idrisi and ENVI, and ESRI's ArcGIS. You may also be interested in the Geospatial Modelling Environment at www.spatialecology.com.

Figure 4.7 illustrates an example of random sampling being used for bulk density and soil moisture content for the $60 \times 60\,m^2$ patch of the Karoo National Park in South Africa illustrated earlier. As for the systematic sample procedure discussed previously, sample locations are illustrated against a backdrop of the full gridded data set. As you will see from the descriptive statistics surrounding each sample, the larger sample better estimates the known data for the population. However, in comparison with the systematic method, the simple random method is less able to determine the characteristics of the underlying population given similar data volumes to those used in the systematic sample.

Looking at equivalent graphs for sample statistics with respect to sample size as those we reviewed earlier (Figures 4.8 and 4.9), we can see that the random sampling method is slow to stabilise. Not until the sample of 380 points do both the average and standard deviation of the sample approach that of the overall population. As before, we see considerably more fluctuation in the standard deviation of the sample from that of the overall grid in the variable of highest standard deviation in its population, bulk density. Simple random sampling carries inefficiencies relative to systematic samples with or without a random element.

As Pearson and Rose (2001) demonstrate in their study estimating the magnitude and distribution of radioactive caesium in a Tennessee reservoir, the relative inefficiency of simple random sampling as a method is not an exception for spatial studies. Indeed, in his book *Statistical Rules of Thumb*, van Belle (2002, p.xx) generalises the situation and suggests that you should 'always consider alternatives to simple random sampling for a potential increase in efficiency, lower costs and validity'.

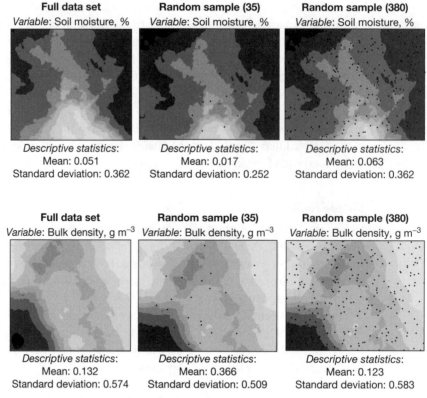

Figure 4.7 Random sampling approach for standardised soil moisture (%) and soil bulk density (g m^{-3}) across a 60×60 m^2 sampling frame in the Karoo National Park, South Africa

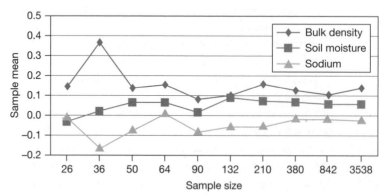

Figure 4.8 Variation of sample mean created using a simple random methodology, by sample size, for variables bulk density, soil moisture and sodium

Prima facie, simple random sampling meets the probabilistic requirements for a representative sample. In reality, should you suspect that sampling units are highly heterogeneous in an unevenly distributed manner, studies such as those of Wolcott and Church (1991) and Gurnell *et al.* (2008) concerning river sediment, or that of Pearson and Rose (2001) concerning reservoir sediment, suggest that

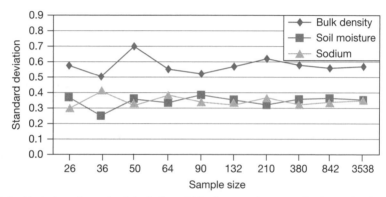

Figure 4.9 Variation of standard deviation within the sample created using a simple random methodology, by sample size, for variables bulk density, soil moisture and sodium

you should consider an alternative method such as stratified random sampling (below). There is also a danger, illustrated in Figure 4.7 for soil moisture in particular, that random sampling can miss capturing small spatial structures unless the sample is particularly large, owing to the uneven distribution of the measured values.

Stratified random

To achieve representative data in a geographical setting, the stratified random sample is particularly important. This is especially so where the population units are more similar within each stratum (for example, land cover, elevation class, geology) than they are across the strata. Under a **proportionate stratified sampling scheme,** the number of units drawn from each stratum is proportionate to the population of the underlying strata. This provides a sample representative of the overall population. Wider availability of digital coverages for soil, geology, elevation and census units amongst others make the use of GIS as part of the sampling design process something worth considering.

Where strata form the basis for the comparative study, for example the variation of a process by geology, you require data that has been selected randomly in all respects *apart from* underlying geology to provide an appropriate comparison. The sampling frame, or area, would be divided into areas according to geology and, under a **disproportionate stratified sampling scheme,** identical numbers of samples would be sampled randomly from each area to provide data that were not biased to one geological type over another. In a similar manner, in their study of newly deposited sand and sediment along an island-braided gravel-bed river channel, Gurnell *et al.* (2008) stratify their sample according to discrete sediment patches representative of particular geomorphological settings. In this case, equal-numbered samples were collected from the surface of established islands or floodplain, the lee of pioneer islands on gravel bar surfaces and open gravel bar surfaces. Rice (2003) also highlights the relative general efficiency of collecting an equal number of random units per stratum and weighting the statistics gathered according to the stratum's size within the overall target population in order to generate appropriate estimates for the population as a whole. The process of weighting the data from the two strata is critical here if representative inference is to be made.

In a human geography setting, you might use statistics derived from the census to select similar but mutually exclusive wards within which to conduct interviews to randomly selected households. This allows sub-groups within the population, important to the study, to gain fair representation. Stratification also might usefully be undertaken by gender or age categories. Be aware too that stratification might usefully be carried out over time rather than space in a longer running study; for example, exposure to particulate matter such as PM_4 has a strongly seasonal component (Fanshawe *et al.* 2008). Stratification by depth may also be relevant in a bio-geographical or macro-geomorphological setting.

There are similarities between quota sampling (described above) and stratified sampling. However, the critical difference is that the selection of individuals (be these points or people) is not truly random for quota sampling and researcher bias can therefore potentially affect selections. For this reason, stratified approaches carry less risk of overgeneralisation. They also carry potential administrative benefits. For example, in an environmental study, the number of agencies or rangers that you need to contact to discuss access to your target population is potentially reduced. Overall, stratified random sampling is a commonly used and effective method for sampling geographical phenomena and is regularly used within academic research. It does, however, rely on accurate information about the processes affecting the phenomenon you are researching and the availability of adequate information concerning appropriate strata.

Multi-stage random

In this method, the sampling frame is divided into multiple hierarchical levels. Each level is sampled randomly in order to select the elements to be included in the next level. Rice (2003) suggests that these methods can be effective when the target population is very large in a temporal or spatial sense. If we were to conduct the undergraduate student alcohol consumption worldwide for example, we might select a subset of countries, then a subset of universities and then, finally, a subset of students from this set to interview. The multi-stage random approach is also useful for ensuring that a full range of scale separations is present in the sampling.

Clustered random

This method is identical to the multi-stage random method, except for the fact that all elements are included at the final (most detailed) level. The method is most commonly used when access to the complete sampling frame is unavailable. We might for example randomly pick six areas from a gridded geographical sample frame and then comprehensively survey each of these for evidence of a rare plant or animal. The critical issue here is that the sample members are chosen as a group (or cluster) rather than as individuals.

There is a problem that can arise in geographical studies when using cluster sampling: that of geographical autocorrelation (for example, Menezes and Tawn 2008). We will come back to this term in more detail when we discuss spatial association in

Chapter 8, but the critical issue is that the value of geographical data is often related to the value of other data in a similar location. In our bio-geographical example here, this could relate to the spread of seed by local agents, for example. An assumption of inferential statistics is that the data are independent and this is a scenario violated by many locally operating geographical processes. In fact, there are adaptations of standard inferential statistics to get round this troublesome problem for geographers (for example, Cliff and Ord 1975). Meanwhile, weigh up the likelihood of this problem for your particular study. If you wish to use inferential methods and have some doubt regarding the independence of your data, this is a sampling method that can compound the problem of spatial autocorrelation and is best avoided. If your purpose is to explore autocorrelation in your data, however, perhaps as part of a pilot study, it can be very useful as a sample method.

Sample size, precision and confidence

We have seen already from the examples for the Karoo that, as a general rule of thumb, a larger sample provides an estimate closer to the population mean but that the benefit of ever-increasing volumes of data starts to drop the larger the sample becomes. In a time-effective study, you should only seek to collect the amount of the data that you need to solve your problem and no more. In seeking this 'optimal' sample size, note that it is both embarrassing and expensive if you find that you need to return to a study at a later date because you have collected too few data and have inconclusive results. Statistical tests also carry minimum sampling requirements, depending on the number of variables you are investigating via your sample.

The difficulty we have is that there is no set sample size that covers all eventualities since, as we saw in regard to our example of soil characteristics in the Karoo, sample size is connected to the underlying **variability** of the phenomenon you are measuring. Only a few measurements might be needed to estimate the mean elevation of a large flat plain, but an area of complex topography would require considerably more. Further, the answer to the number of samples also depends on the desired degree of **precision** you require of your answer. All things being equal, you would need more samples to estimate the elevation of your area to within ±10 cm than ±100 cm. Finally, the third consideration as regards sample size is the **confidence** you wish to achieve in your result. There is a trade-off between precision and confidence given a particular sample size, since it is easier to be certain about a 'woollier' result. We will look more closely later at the way that the term confidence is used in statistics in Chapter 5 (Section 5.8). Chapter 5 will also look at the relationship between sample and population in more mathematical detail. More broadly considered, this section is about data and research quality objectives.

Approaches to managing the sample size issue fall into two camps, the pilot study and the monitoring approach, of which the pilot study is the preferred method where time and project location/scope allow.

The pilot study

Pilot studies are useful to estimate the general characteristics and scale of variation in a phenomenon for which little initial information is to hand before your research starts.

This inherently involves a two-stage sampling process in which information, such as an initial estimate of standard deviation, from the initial sampling phase is used as input to other formulae. This can help us to determine appropriate sample sizes for estimating population values with a certain degree of confidence (Key concept 5.4) in a rather more scientific fashion than the monitoring approach below. Note that, because small samples are inherently biased, estimating the effect of size on the results is imprecise; methods for estimating sample size from a pilot study are the subject of some discussion on topics such as **non-centrality** (for example, Cumming and Finch 2001) which in practical terms go beyond the scope of this book.

Monitoring approach

Using this method, you should monitor your sample characteristics as we discuss below and stop sampling when it is clear that additional measurements are not adding to the data as regards their overall mean and variability. Assuming that you have considered controlling factors and scaling issues, when mean and variability begin to level this is a sign that sufficient data have been collected for the sample to be considered representative.

Key concept 4.10 Sampling error and sample size

As sample size increases, sampling error decreases. More homogeneous populations will have a lower sample error, given the same sample size, than a more variable population.

If you are able to measure your data directly within the field, an empirical rule of thumb is to stop sampling when the standard deviation of your measured values achieves a degree of stability. For indirect measurements, a two-stage sampling process is recommended.

Another common-sense approach to determining how many measurements you may need is to draw on the experiences of a published study using a similar population to yours.

Practically speaking, you should factor some spare time into your fieldwork plan as a contingency to allow additional sample points to be collected as part of the main survey, rather than risk needing to return at a later date. Some authors suggest as much as 20% of the field time should be reserved for contingencies (de Gruijter 1999). Of course, the method can also suggest an earlier finish to your data plan, or that an expansion of geographical extent is feasible.

The issue of sample size is particularly pertinent where the data form the basis for inferential statistical approaches (Chapter 5), in which case you are essentially assuming that your data samples are representative of the population.

An approach that simply adopts a commonly held rule of thumb that a sample size should be at least 30 in regard to one geographical population, or that allows you to stop sampling once the sample mean and variability have steadied, assumes a focused study of limited scope. The sample size gained will not necessarily be sufficient to guard against statistical error. Statistical errors can be classified into two distinct types. Type I error arises when the results of your work suggest that you have proven that a significant relationship between variables exists, when in fact this is not true. In contrast, Type II errors occur when you fail to observe a difference or pattern when in fact one does exist. There is a trade-off between these two types of error; if you only collect a small sample and wish to be highly confident that you will not make a Type I error, then you run the danger of making a Type II error. We will come back to these ideas in Chapter 5. Here, it is sufficient to note that the minimum theoretical sample size that you might need is not necessarily sufficient for a robust research design.

It is important therefore to keep an eye on the developing characteristics of your sample as you collect it, and adopt simple empirical methods to assist you in establishing whether your sample size is sufficient for your purpose where possible. Keeping a running track of the standard deviation and mean in your sample data is good practice. At the point at which the variability in your data stabilises, and assuming that your data collected to that point are representative of the whole as opposed to one spatial or temporal subsection, it may be reasonable to stop collecting further data. The practice of monitoring the changing number of species as a quadrat size is increased until a stable number suggests an optimal quadrat size is one commonly used example of this generic approach. Entering and analysing your data on a laptop, tablet or PC on a regular basis during your field campaign also affords backup for your sample data should a notebook or PDA device go astray or fail.

Key concept 4.11 Sample size and replicates

Sample size usually refers to the number of measurements (units, elements) made from each population.

If you are comparing what you suspect are two different populations, then the term **replicates** is more commonly used, particularly in biological circles.

Finally, it is important to re-emphasise before moving on that there can be *exploratory* value in the results from a small sample that is not representative of a large population. This is particularly the case when considering qualitative reportage that values individual human experience or unusual scientific oddities. However, for quantitative work for which you intend to draw general, statistically robust inference, the gathering of sufficient and representative data is a prerequisite. Regardless of your philosophical approach to geographical research, clarity as to the method applied to the selection of individual cases should always accompany any results such that the generality of the outcomes is not overstated. It is up to you to convince your audience that your sampling strategy was strong enough to justify your interpretations.

4.5   Sampling methods: issues and practicalities

Missing data and personal safety

Your sampling design suggests locations at which to collect data. However, it is inevitable that **missing data** will arise within your sample. This can arise for a variety of reasons, from no response to a postal survey or at a closed door, to damage to an instrument or loss of a small sample of records. In all cases, it is wise to assume that you will have missing data and compensate by selecting a larger initial sample.

You will also need to weigh up the safety aspects of the locations where your design suggests you should collect data. A recent BBC documentary on biodiversity in Guyana saw world-class climbers scale tabletop mountains to investigate niche diversity, but this would not be a feasible study for most of us. Where a sample looks as if it might be compromised for health, safety or accessibility reasons, the volume of such samples should be reviewed. Their absence might compromise the study to the extent that it should be altered to a large degree, or altered in scope such that risky activity is avoided. Climbing unstable scree slopes as a necessity would indicate a change of project question is needed unless acting as part of an experienced climbing team; avoidance of one particular cliff location (of several and with viable alternatives) owing to an unstable overhang would not. In the UK, British Safety Standard BS 8848 (RGS-IBG 2010) provides a systematic overview on how best to manage fieldwork risks of this type. Dismissing one sample point of many because the location could take a whole day to reach could be valid if the other data were well spaced out and isolation (or its effects) were not material to your research question. In a human geography context, you will need to weigh up your physical safety when conducting interviews or questionnaires by location and time of day, and also consider both your and a research participant's emotional safety. Always place your safety first, but always think twice before amending a sample for reasons of convenience alone.

University departments like other organisations will have protocols for both risk assessment and research ethics. Issues associated with a student study should be discussed with a dissertation supervisor as an integral part of the sampling design and scope of the study. Research ethics are not only a matter for human geographers to consider; physical geographers and environmental scientists should also review the environmental ethics of their proposed sampling design.

Data collection: one step in a chain

In many circumstances, you will have the available instrumentation available in the field to measure your sample data directly. However, particularly in some areas of physical geography, you may have to analyse materials further in the laboratory to extract the information required for your study. This raises a number of potential practical and ethical issues that you should consider before launching into data collection.

The purpose of this book is largely to equip you with numerical and statistical expertise for your studies, and laboratory methods and processing are important matters in their own right that require detailed treatment according to the particular domain of your study. However, precisely *because* statistical and laboratory methods are usually treated separately in textbooks, there is the danger that a number of practical considerations linking the two themes fall by the wayside when considering a sampling campaign. Consider for example the following issues, some of which relate particularly to study overseas and others that are more generic:

- *How heavy is your sample?* If you are far from home, can you afford to transport your samples back to your laboratory or could you arrange to analyse them overseas? The cost of freighting rock samples back from another continent can run into thousands of pounds.

- *Import/export restrictions.* Pest risk assessment regulations in many countries restrict the movement of organic matter. You may need to consider using cruder technical approaches to extract the information you require that can be conducted *in situ,* or analyse your sample in the country using a reciprocal arrangement between research laboratories or by working as part of an international team.

- *Biological capital.* The removal of environmental materials from one nation to another is a sensitive issue on ethical grounds.

- *Analytical costs.* The laboratory analysis of samples can be expensive in terms of both the costs of chemicals required for processing and the processing time. Samples need to provide as much information as possible. Review the scope of your study and sampling strategy with these temporal and monetary factors in mind.

- *Volume of sample.* How much material is required to conduct subsequent chemical analysis? If you wish to conduct multiple analyses for each sample, make sure that the volume of material that you collect is sufficient. If you are trying to answer multiple questions in your study, each measurement will add a *degree of freedom* to your study, which further means that higher volumes of data need to be collected. Degrees of freedom are discussed in more detail in Chapter 2 (Key concept 2.1).

This is not intended as an exhaustive list of practical matters to consider when you are intending to process sample material in the laboratory subsequent to a field campaign, but rather to highlight potential logistical dangers that can catch you unawares. We recommend that you discuss your sampling campaign from both statistical and practical perspectives with an expert in the subject before committing yourself to a particular plan.

Instrument accuracy and scale

In Section 4.3 above we considered the matter of scale in regard to sampling design. Here, we highlight the need to consider the accuracy of your measurement instrument as part of your scale review. The purpose of this book is not to review the many forms of specialist instrumentation increasingly being used in many sub-domains of geography and environmental science, but consideration of these matters does form an important element within the methodological design process that links with sampling. The best way to explain this matter is by example, and we illustrate the matter here using global positioning systems (GPS) since they are familiar to many and are commonly used in a variety of research projects.

Global positioning systems are now widely used across geographical research to support the recording of locational information for a wide variety of subjects. However, the accuracy and precision with which GPS record location and indeed elevation varies according to a variety of factors relating to the environment that is being measured and the sensitivity of the instrument itself. Taking variation in the environment first, errors in position occur where the clear view from GPS to satellite(s) is blocked, for example by buildings or trees. When the number of satellites visible to a receiver drops down to three or less, the GPS is unable to give a position without the use of aiding techniques (Li *et al.* 2005). Apart from satellite visibility, the receiver–satellite geometry also affects the accuracy of GPS. Hence, particularly if you plan to use GPS within a tropical forest area or a dense urban area such as Manhattan in New York for your fieldwork, make sure that you keep an eye out for the number of available satellites and the values of Precision Dilution of Position (PDOP) shown on your GPS screen or within the GPS file record (an NMEA file) that you subsequently download. In some cases, a grid reference provided by 1:10 000 government agency mapping (where available) will be more fit for purpose than a GPS reading *where a detailed locational record is a critical element of the study*. In many parts of the world, access to 1:10 000 mapping is an unaffordable or unobtainable luxury, in which case a GPS location plus understanding of its associated error remain valuable data. Accuracy and precision are context dependent, as well as purpose related.

This issue of purpose, or fitness for use, is vital when weighing up sample precision. If you are using GPS as a record of your field days and for general photograph tagging, a relatively crude locational trail and note of well-spread field site locations (for example, pebble size at different points along several kilometres of the reach of a river) will be of value. A handheld GPS with a common Sirf III chip (accuracies within 4 m, but commonly much better) could handle this scenario and could also be used to record the location of bio-geographical samples across a 500 m to 1 km transect in an open area. Using such a device to measure locations along a 10 m transect would, however, be inappropriate given the ratio of potential error to distance measured. Similarly, a human geographer might use a handheld device to tag photographs or audio reflections of participants in a study across different areas of the city, or use GPS to tag photographs of urban graffiti.

If, however, you are collecting ground data for use in a satellite classification study and your remotely sensed image pixel size is 20 m but your GPS accuracy is above half this size in an area of variable land cover, then sample precision would be compromised. Assuming that high-precision base mapping data are not available as an alternative, it would be better to seek a more consistent example of the vegetation you wish to classify and locate your sample there. If tightly defined samples are required, as is commonly the case in remote sensing research, the use of a more precise instrument such as differential GPS together with postprocessing of output should be considered.

As a general rule, studies using GPS to record elevation across local areas for use in numerical analysis work should be avoided unless you have access to differential GPS equipment and training in the postprocessing of data. It might seem tempting to use handheld single unit GPS to estimate cliff height, for example. If, however, you consider the strong likelihood of a 10 m error in both upper and lower measurements, the overall error as a portion of cliff height will be high. It is very often the case that the nearest contour will provide a better estimation of elevation than single GPS.

> **Key concept 4.12 Measurement accuracy**
>
> It is essential that you consider sample measurement accuracy as a component of sample design and ensure that your data collection strategy is fit for purpose as regards precision and scale. Measurement error can be caused by both instrument and observer.

Whilst this discussion has focused on GPS in particular, these issues of measurement precision and scale arise in many other aspects of geographical research. Similar caveats relate to the accuracy and regulatory ability of your temperature sensor or size, orientation and internal variability of a vegetation quadrat, among other items. In the case of many different types of instrument used to measure chemical or physical properties in particular, you should calibrate the instrument you are using carefully to avoid systematic error. Further, repeated measurements of the same phenomenon are common practice to guard against measurement error and hence sample bias in many areas of physical geography. Sampling protocols are often suggested by manufacturers and also exist in many specialist domains of interest to geographers. The critical issues, regardless of the variable in question, are to research instrument and local site protocols relevant to your study, weigh up measurement error against phenomenon size and expected variation, and include such considerations as part of your sampling and data collection plan. Further, you should note the actions you took to avoid measurement or local observation bias as part of the metadata associated with your particular study.

Key points

The main steps in the overall sampling process are summarised in Figure 4.1, which draws together both general and mathematical considerations related to sampling. Critically, in regard to your sampling campaign, the following points are key:

- Having a clear research question and focused scope for a study are vital prior to considering sample design.
- It is neither feasible nor indeed necessary to measure or interview an entire population; a sample should be gathered.
- Gathering a representative sample, free of bias, is critical if you are planning to make statistical inference from your sample.
- A focused representative study will be backed up by reading that establishes controlling factors affecting the phenomenon or process you are studying.
- If you wish to use inferential statistics (see Chapter 5) as part of your study, you should adopt a probabilistic sample method (for example, simple random, systematic or systematic–random sampling).

- Systematic–random sampling is an effective and efficient form of sampling for quantitative geographical research.

- Sample design needs to be considered from both statistical and practical perspectives; issues relating to the data processing chain and measurement accuracy can impact on the subsequent usability of the data.

- Your sample should be sufficiently large to be able to draw inference. Keep a watching brief on the variability of your data as you collect them, and consider a two-stage study if there is little literature discussing similar samples to those you plan to collect.

References

Bromley, R.D.F. and Thomas, C. (1999) The geography of shoplifting in a British city: evidence from Cardiff. *Geoforum,* 27, 409–423.

Cliff, A. and Ord, J.K. (1975) The comparison of means when samples consist of spatially autocorrelated observations. *Environment Planning A,* 7, 725–734.

Cox, D.R. (1952) Estimation by double sampling. *Biometrika,* 39, 217–227.

Cumming, G. and Finch, S. (2001) A primer on the understanding, use and calculation of confidence intervals that are based on central and non-central distributions. *Educational and Psychological Measurement,* 61, 532–574.

de Gruijter, J. (1999) Spatial sampling schemes for remote sensing. In A. Stein (ed.), *Spatial Statistics for Remote Sensing,* Dordrecht: Kluwer Academic, Chapter 13, pp. 211–242.

Dickie, J. (2007) Relationships among the physical and chemical properties of soil, vegetation and land degradation in semi-arid environments. PhD thesis, University of Leicester.

Durrell, G. (1954) *Three Singles to Adventure,* Harmondsworth: Penguin, p.217.

Fanshawe, T.R., Diggle, P.J., Rushton, S., Sanderson, R., Lurz, P.W.W., Glinianaia, S.V., Pearce, M.S., Parker, L., Charlton, M. and Pless-Mulloli, T. (2008) Modelling spatio-temporal variation in exposure to particulate matter: a two-stage approach. *Environmetrics,* 19, 549–566.

Gurnell, A.M., Blackall, T.D. and Petts, G.E. (2008) Characteristics of freshly deposited sand and finer sediments along an island-braided, gravel river bed: the role of water, wind and trees. *Geomorphology,* 99, 254–269.

Jarvis, C.H. (2000) Insect phenology: a geographical perspective. PhD thesis, University of Edinburgh.

Li, J., Taylor, G. and Kidner, D. (2005) Accuracy and reliability of map matched GPS coordinates: dependence on terrain model resolution and interpolation algorithm, *Computers and Geosciences,* 31, 241–251.

McAfee, R.P. (2002) *Competitive Solutions: The Strategist's Toolkit,* Princeton University Press: Princeton, NJ.

Menezes, R. and Tawn, J. (2008) Assessing the effect of clustered and biased-multistage sampling. *Environmetrics,* 20, 445–459.

Met Office (2010) Observations. National Meterological Library and Archive Fact Sheet 17 – Weather observations over land. Available at http://www.metoffice.gov.uk/corporate/library/factsheets/factsheet17.pdf, accessed 21 October 2010.

Parsons, A.J. (1982) Slope profile variability in first-order drainage basins. *Earth Surface Processes and Landforms,* 7, 71–78.

Pearson, S.M. and Rose, K.A. (2001) The effects of sampling design on estimating the magnitude and distribution of contaminated sediments in a large reservoir. *Environmetrics,* 12, 81–102.

RGS-IBG (2010) BS 8848 a new British Standard. Available at http://www.rgs.org/OurWork/ Advocacy+and+Policy/Outdoor+learning+and+fieldwork+policy/BS+8848+a+new+British+ Standard.htm, accessed 21 October 2010.

Rice, S.P. (2003) Sampling in geography. In N.J. Clifford and G. Valentine (eds), *Key Methods in Geography,* Sage: London, pp. 223–247.

Smith, M.J., Rose, J. and Booth, S. (2006) Geomorphological mapping of glacial landforms from remotely sensed data: an evaluation of the principal data sources and an assessment of their quality. *Geomorphology, 76,* 148–165.

Smith, T.M.F. (1996) Public opinion polls: The UK General Election 1992. *Journal of the Royal Statistical Society Series A, Statistics in Society, 159,* 535–545.

van Belle, G. (2002) *Statistical Rules of Thumb,* Hoboken, NJ: Wiley.

Wolcott, J. and Church, M. (1991) Strategies for sampling spatially heterogeneous phenomena: the example of river gravels. *Journal of Sedimentary Petrology, 61,* 534–543.

From description to inference

Chapter overview

We previously have indentified three uses for statistics. These are description, inference and explanation. To this point, the focus has been on description, summarising and showing the centre, spread and shape of a data set. However, Chapter 3 moved towards inference: looking beyond the data in their own right and using them to gather information about the properties of the system, process or structure that has been measured.

In this chapter we build on the important distinction between a sample and its population that was introduced in Chapter 4. Because it is rarely possible to survey a population in its entirety, we aim to have a representative sample of if. However, this creates a problem: if what we learn about a population is based on what we find in the sample, then our information must be sample dependent. Changing the sample would change what we learn about the population.

This chapter looks at how we can use the mean and the standard deviation of the sample to estimate the same for the (entire) population. Although we rarely know the true mean of the population, we can calculate the probability that it falls within a specific range of values. We can do this because of the central limit theorem and because of the properties of a normal curve that we reviewed in Chapter 3.

Learning objectives

By the end of this chapter you will be able to:

- Explain the difference between describing a data set in its own right and using the data to infer properties of an underlying population.

- Understand why it is rarely possible to enumerate a population in its entirety so sampling methods are used instead.

- Emphasise the distinction between a sample and a population, and the notation that is used to distinguish them.

- Appreciate that a sample-based estimate of a population mean is dependent on the sample taken.

- Recognise that there are as many possible sample means as there are ways of sampling the population, but we expect some values to be more probable than others.

- Understand that the standard error is a measure of unreliability that increases with the variation in the data but decreases with the square root of the sample size.

- Explain why we do not need to know the true value of the population mean to be '95% confident' that it is within 1.96 standard errors of the sample mean, and '99% confident' that it is within 2.58 standard errors.

- Know that for small samples we use a slightly different method to calculate confidence intervals and that this is based on the properties of the *t*-distribution instead of those of the normal curve.

5.1 About inference

Inference is at the heart of how and why statistics developed. William Gosset (1876–1937), for example, was employed at the Guinness brewery in Dublin (and later in London) where he developed experimental trials and statistical analysis to test ways of improving barley yields. Because of the commercial applications of his work, Gosset was required by his employers to publish under a pseudonym. He chose Student (see Boland 1984; Salsburg 2002).

The challenge of an agricultural trial is to determine whether the effect of some 'treatment' or intervention such as using fertiliser has the desired outcome. For Guinness this would be a greater barley yield without impairing the taste of the brew.

The basic way to undertake the experiment is to try the treatment on some plots of land but not on others and then determine whether the intervention produces a higher yield on average. However, simply detecting an increase is inadequate proof of success. What also needs to be considered is whether the treatment *consistently* produces a higher yield. Therefore, the spread of values around the average is looked at too.

Even so, it is always possible that the increase is not actually due to the treatment but is caused by some other hidden or unexplained factor. The skill is to minimise that possibility by careful experimental design and sampling (see Chapter 4). Having done so, the analyst may determine the probability of failure – of no increase in yield – if the treatment were applied again in similar circumstances. If that probability is low, then, reciprocally, the probability of success is high, and the analyst may infer that the treatment causes increased yield. There is reasonable proof that it does. Yet, it remains an inference: it is a conclusion based on what has been measured and observed. There is always a chance that it is wrong.

We gave another example of inference at the beginning of the book. There we suggested that 'students tend to be uneasy about statistics' whilst admitting the sentence is a generalisation. We did not claim the statement to be definitively true because we could not – we have not met every student to validate it. Instead, we were demonstrating a principle:

> to learn from a sample (of students), something we have confidence will be generally true for other members of the (student) population too, whilst at the same time being mindful of variation (the fact that some students like statistics more than others).

The properties of the sample will determine how successful the inference will be. The sample should be representative of the population for which 'it talks' and not inappropriately biased towards particular sub-groups.

If you were interested in general social attitudes to employment and childcare, for example, you might obtain misleading results were you to sample only males, or members of just one cultural/ethnic group. The sample would be inaccurate.

5.2 A measure of unreliability

Assuming the sample is indeed representative, then what other factors would lead to reliable inference?

Crawley (2005) tackles the question from an opposite direction – in terms of a measure of *unreliability*.

All other things being equal, would you be more suspicious of our drawing conclusions from a sample of 1000 students or from a sample of 10 students? We would worry about the latter. It contains less information upon which to hang our inferences. **Unreliability increases as the sample size decreases**.

However, we need also to consider the variation in the data. If 95% of students surveyed rate their liking of statistics to be low, then this implies greater consensus within the population than if half dislike statistics whilst the other half love them! Put another way, variation in the sample suggests variation in the population. We can say that **unreliability increases with variability**, and we know that variability can be expressed as the standard deviation of a set of measurements.

Expressing these ideas in terms of an equation gives

$$\text{unreliability} = \frac{s}{\sqrt{n}} \qquad (5.1)$$

where *s* is the standard deviation and *n* is the sample size. The presence of the square root means that the cost of collecting extra data must be balanced against the diminishing return for doing so. For example, whereas a sample size of $n = 10$ might be 10 times more unreliable than a sample of $n = 1000$ (with 990 more respondents), to improve the reliability 10-fold would require a sample of $n = 100\,000$ (an increase of $99\,000$).

5.3 What is a population?

The word sample means a selection from (or subset of) a population. Unfortunately, what is meant by population can be more abstract and harder to conceive. For, say, an electoral survey of voting intentions, the population is easy to imagine: it is every registered voter. However, what is the population if we are interested in the effects of a fertiliser on average crop yield?

One definition of population is 'every possible object (or entity) from which the sample is selected'. Another way of looking at it is as 'the complete set of all possible measurements that might hypothetically be recorded' for the study. This is how the (target) population was defined in Chapter 4, Key concept 4.1.

For an agricultural trial the population is every possible way the land could be divided into plots to determine the average yield. In other words, every average that is potentially calculable. This is theoretical because the land can be subdivided in an infinite number of ways, even if the number of plots is held constant. A new sub-division is achieved simply by moving the boundaries of the plots very slightly. The population is infinite because what we are measuring is part of a continuous field, not a finite set of discrete objects.

Despite being abstract and somewhat intangible, it is useful to understand the conceptual difference between a sample of data (a set of measurements) and the population from which the sample is drawn (what, ultimately, we are measuring).

As in the example above, populations rarely are completely measurable; there are too many measurements that could be taken. Even if they could all be taken, the cost of doing so would be prohibitive (and not really necessary, because of the 'diminishing returns' we discussed after Equation 5.1; see also the discussion in Chapter 4, Section 4.4). Our way into understanding properties of the population – properties such as its mean and standard deviation – is to use what we can find out from the sample.

Knowledge of the population will remain incomplete. What we infer will be a best guess but an informed one, based on the measure of unreliability and the properties of the normal curve discussed in Chapter 4.

Using notation to distinguish a sample from a population

As you read this and later chapters, **have in mind the distinction between the sample and its population**. Recognising this distinction is the key to understanding inferential statistics and will help avoid confusion. Understand that the statistical tests are trying to go beyond the data themselves to say something about the population from which they are drawn.

The distinction brings some extra notation. We continue to use $\bar{x}$ to indicate the mean of a sample of data and now introduce μ (pronounced 'mew') to be the mean of its population. The standard deviation of the sample is still indicated by s, whereas the standard deviation of the population will be σ (pronounced 'sigma').

When we cannot observe a population in its entirety (which is usual) then its true mean and standard deviation will remain unknown. However, we can use the mean and standard deviation of the sample to provide estimates. In notation, we will use $\bar{x}$ and s to provide estimates of μ and σ. The rest of the chapter is about how we do so and about our confidence in estimating μ.

Key concept 5.1 Inference, samples and populations

Inference is at the heart of statistics. A general principle of inferential statistics is to determine, from a sample, the probability that some characteristic of the population suggested by the sample is true. This requires the sample to be representative of the population for which 'it talks'.

The word **sample** means 'a selection from (or subset of) a population'. The population can be more abstract. One definition of **population** is every possible object (or entity) from which the sample is selected. Another is the complete set of all possible measurements that might hypothetically be recorded (cf. Key concept 4.1).

The distinction between the sample and population is important, giving rise to pairs of notation. Where $\bar{x}$ is the mean of the sample, μ is the mean of its population. The sample standard deviation is s; the standard deviation of the population is σ.

Often μ and σ are unknown and unknowable. However, the mean and standard deviation of the sample help provide estimates of them.

 ## 5.4 A geographical example

Imagine you are a surveyor interested in determining average height above sea level of the land illustrated in Figure 5.1, assessing its susceptibility to rising sea levels. Elevation is continuous, it changes in all directions and in all places albeit sometimes imperceptibly. Since it is impossible to measure at an infinite number of locations, samples must be taken and their mean calculated.

For this example, the samples are drawn in two ways giving two batches of data. The first – the crosses in Figure 5.1 – are taken at regular intervals (a systematic sample). The second, the circles, are randomly located. In Table 5.1, the first sample is A and the second is B. The table includes the measure of unreliability introduced in Section 5.2. Figure 5.2 compares the samples using box plots.

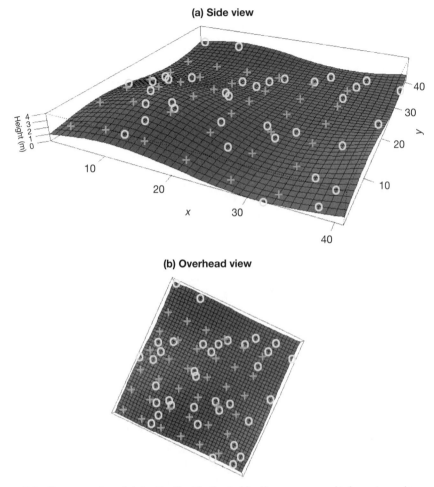

(a) Side view

(b) Overhead view

Figure 5.1   Two samples of data, the first indicated by the crosses and taken at regular intervals across the study scene, the second indicated by the circles at randomly chosen locations

Looking at either the table or the box plots, the two samples of data evidently are not the same. Comparing the mean of batch A ($\bar{x}_A = 1.106$ m) with the mean of batch B ($\bar{x}_B = 0.893$ m), they are found to differ. Since both provide estimates of the 'true' mean height of the study region – the population mean – what we can immediately conclude is that our estimates of it, μ, are dependent upon the sample – upon when, where and how it was taken.

Table 5.1   Summary of the two batches of height data. Batch A is the crosses in Figure 5.1, batch B is the circles. The heights are in metres

Batch	n	Min.	Q1	Median, $\tilde{x}$	Mean, $\bar{x}$	Q3	Max.	s	'Unreliability'
A	36	0.002	0.089	0.359	1.106	1.951	3.807	1.226	$1.226/\sqrt{36} = 0.204$
B	36	0.000	0.097	0.572	0.893	1.081	3.807	1.029	$1.029/\sqrt{36} = 0.172$

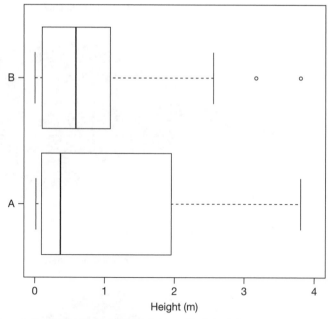

Figure 5.2 Box plots comparing the two batches of height data

5.5 A thought experiment

Imagine we could sample the study scene in Figure 5.1 in every way possible, holding constant the size of the sample (*n*). We cannot but we can approximate the process using a method of simulation to sample the landscape a very large number of times. This will give us much more complete information about the population than would usually be possible by sampling. Of course, we can only do so because the landscape shown in Figure 5.1 is a computer model; it is not real. The purpose of the exercise is demonstrative: what would happen if ... ? We are not suggesting it could be repeated in the field.

The simulation proceeds as follows. First, we use a computer to create a new set of *n* = 36 sample points, each randomly located within the scene. We then read off the height value at each location. This creates a third batch of data in exactly the same way as for batch B but sampling at different locations. Next, we calculate the mean of those heights. Having done so, we repeat the process of sampling and of averaging a further 9997 times. This gives us 10 000 batches of data in total, 10 000 sample means and therefore 10 000 estimates of the landscape's true mean height.

We already know that $\bar{x}_A = 1.106$ m and that $\bar{x}_B = 0.893$ m. We can now tell you that $\bar{x}_C = 1.028$ m, $\bar{x}_D = 1.038$ m, $\bar{x}_E = 1.071$ m ..., but instead of listing each of the 10 000 sample means separately, we more simply note that some are more unusual than others.

Because the samples are located randomly, every so often they will be taken in peculiar places and generate an atypical result. However, the probability of getting an extremely atypical result is less than the probability of getting a quite atypical result, which is less than getting 'just a little atypical' result. The situation is analogous to the discussion of the height data in Chapter 3 (Section 3.3), where we said that 'exceptionally tall people are rarer than very tall people, very tall people are less common than quite tall people, and quite tall people are less frequently observed than tall-ish people'.

We confirm the analogy by looking at the distribution of the sample means. It is shown using a histogram, a rug plot and also a quantile plot in Figure 5.3. Notably, it is normal. What we are seeing is a demonstration of the central limit theorem introduced in Chapter 3, Key concept 3.3.

Let us be clear about what Figure 5.3 is showing. We have 10 000 sample means 10 000 estimates of the study region's mean elevation and it is their distribution that is displayed in the plots. Some sample means are more usual than others and appear near the centre of the distribution. Others are literally 'far out'.

Intuitively, which provides the better estimate of the landscape's true mean height: the values near the centre of the distribution or those towards the edges? Presumably it is the former because they are more commonplace – they are a safer bet.

We could replace the samples or even change the shape of the landscape and begin the experiment again. We will always achieve the same finding. The central limit theorem suggests the distribution of the sample means will be normal and, critically, *centred on the true mean – the population mean, μ.*

There are four caveats, however. First, if we change the shape of the population then that would change the mean and the standard deviation of the distribution of the sample means. Nevertheless, to reiterate, it remains normal and centred at the population mean.

Secondly, there has to be sufficient variability for the central limit theorem to apply. If the landscape were perfectly flat or, more plausibly, the observations gave only a few discrete values (perhaps because the measuring instrument was not sensitive enough) then the distribution of the sample means would not be normal.

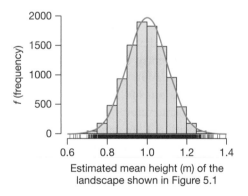 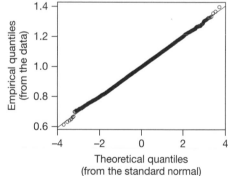

Figure 5.3 A histogram, rug plot and quantile plot showing the distribution of the 10 000 sample means. The distribution of the sample means is normal and is a demonstration of the central limit theorem. See text for details

Thirdly, the sample size cannot be too small. Usually a sample size of 30 or more is considered acceptable for the central limit theorem to be valid (that is, $n \geq 30$). Later in the chapter we will see what happens to the shape if the sample size is less.

Fourthly, remember that we are talking about a distribution that would be seen *if* we had very many samples and were able to compare their sample means. In practice we will not and therefore the distribution is hypothetical and unseen. Nevertheless, the fact that we know what would happen if we had a large number of samples is information that can usefully be applied, as the following section shows.

Key concept 5.2 The distribution of the sample means

A sample mean depends on how, when and where the data are collected.

The **distribution of the sample means** is what we would see *if* we could take every possible sample (of a given size, *n*) from a population, calculate each sample mean, and then plot the distribution of those means using a histogram or quantile plot.

Assuming each sample contains about 30 observations or more (and that there is variability in the sampled data) then the distribution would be normal, because of the central limit theorem (Key concept 3.3). It would be centred upon the population mean, μ.

5.6 Why the thought experiment is useful

You may still be wondering why our thought experiment was useful. It did, after all, rest on the self-contradictory premise of sampling the data an impossibly large number of times.

It is the result that is important – that the set of very many sample means is normally distributed. This tells us that where we have a sample of data and have found the sample mean, that mean lies somewhere along a normal distribution even if we do not know exactly where.

Again, that does not sound especially useful until we turn to what we learned about a normal curve in Chapter 3. We know that 95% of the area under a normal curve is within 1.96 standard deviations of its centre which, for the distribution of the sample means shown in Figure 5.3, is the population mean, μ. Similarly, we know that 99% of the area is within 2.58 standard deviations of the centre.

It follows that if we take a random and unbiased sample, there is a 95-in-100 chance that its sample mean, $\bar{x}$, will be within 1.96 standard deviations of the population mean, μ. We do not need to know what the population mean is, to still know there is a probability of $p = 0.95$ that it is no more than 1.96 standard deviations away from the sample mean. Similarly, there is a probability of $p = 0.99$ that the distance between $\bar{x}$ and μ is 2.58 standard deviations or less.

At the risk of confusion, the standard deviation we are talking about here is the standard deviation of the (hypothetical) distribution of the sample means. This is a bit of a mouthful and it is usually given a different name, the standard error of the sample means, alternatively the standard error of the mean (SEM) or, more lazily, the **standard error** (abbreviated to *se*).

5.7 Calculating the standard error

We now know the probability that the sample and population means are within a certain number of standard errors from each other. We next need to learn how to calculate the standard error.

In principle, the standard error of the mean (SEM) is given using the formula

$$\text{SEM} = \frac{\sigma}{\sqrt{n}} \tag{5.2}$$

(a proof of this can be found in Frank and Althoen 1994, pp.288–289, for example).

However, Equation 5.3 requires us to know the standard deviation of the population, which is indicated by σ. Unfortunately, unless we can enumerate the population in its entirety then σ is unknown. Instead we focus on what we do know, the standard deviation of the sample, s.

Substituting s for σ gives an estimate of the SEM:

$$se_{\bar{x}} = \frac{s}{\sqrt{n}} \tag{5.3}$$

This should be familiar. It is the same measure of uncertainty that began the chapter (see Equation 5.1).

What we can now say is that we do not need to know the value for the unknown population mean to still know there is a probability of $p = 0.95$ that it will be within 1.96 standard errors of the sample mean, and a probability of $p = 0.99$ that it is within 2.58 standard errors, where the estimate of the standard error is $se_{\bar{x}}$:

$$
\begin{aligned}
P(\bar{x} - 1.96se_{\bar{x}} < \mu < \bar{x} + 1.96se_{\bar{x}}) = 0.95 \\
P(\bar{x} - 2.58se_{\bar{x}} < \mu < \bar{x} + 2.58se_{\bar{x}}) = 0.99
\end{aligned} \tag{5.4}
$$

Key concept 5.3 Standard error of the mean

Standard error is a measure of uncertainty. It increases with the variability of the data but decreases as more data are collected. The standard error of the mean (SEM) is estimated as

$$se_{\bar{x}} = \frac{s}{\sqrt{n}}$$

where s is the standard deviation of the data and $\sqrt{n}$ is the square root of the sample size.

5.8 Confidence intervals

Consider, again, that you are a surveyor interested in the mean height of the landscape represented by Figure 5.1. You cannot collect data everywhere so you collect a sample of height values instead. Assume that is sample B in Table 5.1. Those data have a mean of $\bar{x} = 0.893$ m and a standard deviation of $s = 1.029$ m. The sample size is $n = 36$. By Equation 5.3, the standard error is $1.029/\sqrt{36}$ and is equal to 0.172 m. Substituting those values into the first line of Equation 5.4 (and reducing the precision to two decimal places) gives

$$P(\bar{x} - 1.96se_{\bar{x}} < \mu < \bar{x} + 1.96se_{\bar{x}}) = 0.95$$
$$P(0.89 - 1.96 \times 0.17 < \mu < 0.89 + 1.96 \times 0.17) = 0.95$$
$$P(0.89 - 0.33 < \mu < 0.89 + 0.33) = 0.95 \tag{5.5}$$
$$P(0.56 < \mu < 1.22) = 0.95$$

This is called a **confidence interval**. You still do not know the true mean height but you can be '95% confident' that it is within the range from 0.56 to 1.22 m. That level of confidence comes from the probability: $p = 0.95$ (or 95%).

In the same way, the 99% confidence interval is found to range from 0.45 to 1.33 m:

$$P(\bar{x} - 2.58se_{\bar{x}} < \mu < \bar{x} + 2.58se_{\bar{x}}) = 0.99$$
$$P(0.89 - 2.58 \times 0.17 < \mu < 0.89 + 2.58 \times 0.17) = 0.99$$
$$P(0.89 - 0.44 < \mu < 0.89 + 0.44) = 0.99 \tag{5.6}$$
$$P(0.45 < \mu < 1.33) = 0.99$$

It is important to understand what it means to say we are 95% or 99% confident. What it does *not* mean is that 95-in-100 times the population mean is within the interval from 0.56 to 1.22 m, or 99-in-100 times it is within the interval from 0.45 to 1.33 m. This is because the population mean is assumed fixed and constant. It is either within the interval or it is not. It cannot be there on some occasions and not others.

Instead, '95% confident' says that, of all the samples that could have been collected, and of all the sample means that could have an interval of $1.96 \times se_{\bar{x}}$ placed around them, 95% of them will contain the unknown population mean. Similarly, '99% confident' says that, of all the sample means that could have an interval of $2.58 \times se_{\bar{x}}$ placed around them, 99% will contain the unknown population mean. This is linked to the view of probability that asks what would happen in the long run (see the discussion at the end of Chapter 3, Section 3.6).

In principle, a confidence interval could range from negative to positive infinity, thereby including every possible value for the unknown population mean. We would then be 100% confident it contained the population mean.

Such a wide interval is neither useful nor necessary. We know from the properties of a normal curve that although some values can be found in the tails, values are more frequently found closer to the centre. This suggests we do not need to extend the confidence interval out to the extremes for it to be more likely than not it will contain the unknown population mean.

However, unless the confidence interval does contain every possible value then there is always a chance that the population mean will not be within it; 100% confidence is

95% confidence intervals

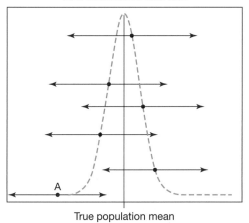

True population mean

99% confidence intervals

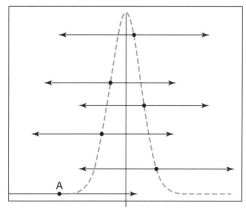

True population mean

Figure 5.4 A selection of sample means from those included in Figure 5.3. The sample means are shown as black dots, the arrows represent the confidence intervals. Observe that the wider the confidence interval is, the more likely it will contain the true but usually unknown population mean: compare the 95% and 99% confidence intervals for sample mean A, for example

pointless – it is not useful to know the population mean lies somewhere within an infinite range of values! The issue therefore becomes about how confident we want to be.

Wider intervals give greater confidence but also produce a less precise estimate of what the population mean could be. For example, the 95% confidence interval for the batch B data suggests $0.56 < \mu < 1.22$. The 99% one is less precise, suggesting $0.45 < \mu < 1.33$. There is a trade-off between confidence and precision.

The level of confidence the analyst wants partly determines the width of a confidence interval. Values of 95% and 99% are conventional, though ultimately arbitrary.

Having made that choice, the width is also determined by the standard deviation and by the size of the data set, n. That follows from the standard error. For a fixed level of confidence, the interval will *widen* with the standard deviation of the sample and *narrow* with the square root of the sample size. This makes sense: the more varied the data are, the more the sample mean could vary from the population mean; the more of the population we sample, the more we know about it, and the more precisely we can predict its mean.

Figure 5.4 illustrates the 95% and 99% confidence intervals for some of the 10 000 sample means included in Figure 5.3. It makes the point that the larger the confidence interval is, the more likely it will include the unknown population mean.

5.9   Consolidation and worked examples

Table 5.2 is a summary of what is known or unknown about a sample and its underlying population. It summarises what can be inferred about the population from the sample, and how.

We use Table 5.2 to find the 95% confidence interval for the batch A data in Table 5.1.

The mean and standard deviation of the data are known. Batch A has a sample mean of $\bar{x} = 1.106$ and a standard deviation of $s = 1.226$ (metres). The sample size is

Table 5.2 Summarising what is known and what can be inferred given an unbiased sample of a population

Mean of the sample, $\bar{x}$	*Known:* $\bar{x} = \dfrac{\sum_{i=1}^{n} x_i}{n}$
Standard deviation of the sample, s	*Known:* $s = \sqrt{\dfrac{\sum_{i=1}^{n}\left(x_i - \bar{x}\right)^2}{n-1}}$
Standard deviation of the population, σ	*Unknown:* We only know s
Where the sample mean lies on the distribution of all possible sample means	*Unknown,* but because of the central limit theorem we know the distribution is normal, centred on the mean of the population and has a standard deviation equal to the SEM
Standard error of the mean (SEM)	Equal to $\sigma/\sqrt{n}$ but this requires the standard deviation of the population to be known. Instead, it is estimated using the standard deviation of the sample, $se_{\bar{x}} = s/\sqrt{n}$
Mean of the population, μ	*Unknown,* but the probability that it is within a given number of standard errors either side of the sample mean can be calculated
	For a '95% confidence interval': $P(\bar{x} - 1.96se_{\bar{x}} < \mu < \bar{x} + 1.96se_{\bar{x}}) = 0.95$
	For a '99% confidence interval': $P(\bar{x} - 2.58se_{\bar{x}} < \mu < \bar{x} + 2.58se_{\bar{x}}) = 0.99$

$n = 36$. The standard error of the mean can be estimated as $s/\sqrt{n}$, which is $1.226/\sqrt{36}$, equal to 0.204.

Substituting the mean and standard error into Equation 5.5 gives (to two decimal places)

$$P(\bar{x} - 1.96se_{\bar{x}} < \mu < \bar{x} + 1.96se_{\bar{x}}) = 0.95$$
$$P(1.11 - 1.96 \times 0.20 < \mu < 1.11 + 1.96 \times 0.20) = 0.95$$
$$P(1.11 - 0.39 < \mu < 1.11 + 0.39) = 0.95 \qquad (5.7)$$
$$P(0.72 < \mu < 1.50) = 0.95$$

Based on *this* sample of data, we are 95% confident that the true mean height of the study region is within the range from 0.72 to 1.50 m.

Use Equation 5.6 now to calculate the 99% confidence interval.

Your answer:

> **Key concept 5.4   Confidence intervals**
>
> **Confidence intervals** define a range of values around the sample mean for which the probability that the population mean falls within it is calculated.
>
> Because of the central limit theorem and the properties of the normal curve (and given continuous data of a reasonable sample size), we can be '95% confident' that the unknown population mean is within 1.96 standard errors either side of the sample mean, and '99% confident' that it is within 2.58 standard errors.

5.10   Finding a 99.9% or any other confidence interval

It would be limiting to restrict confidence intervals to only a 95% or 99% confidence, albeit that these are often chosen, by convention.

Consider a medical trial where the pharmaceutical company is interested in minimising the average number of adverse reactions to a new drug. Before the drug is dispensed more widely, the company and the licensing authority will want to ask, 'how confident are we that the average is not much higher than we think and was only low because of some chance characteristics of the sample?' Dependent upon what is being treated and the nature of any reactions, a very high level of confidence may be sought.

A 99.9% confidence interval is used if the analyst wants to be almost certain that the interval includes the unknown population mean. The general process to determine the confidence interval (CI) is as follows.

First, decide on a level of confidence, here 99.9%. The decision depends upon the context and may be informed by past studies. It may seem desirable always to set the confidence very high. However, in Chapter 6 we will discuss why the decision is not that simple. For now it is sufficient to understand that if we always avoid the risk of mistake we may also avoid discovering a useful (and socially beneficial) result.

Secondly, calculate the probability of 'getting it wrong' – the probability that the confidence interval *does not* include the unknown population mean. This is called the **alpha value** and is equal to

$$\alpha = 1 - (\text{CI}/100) \tag{5.8}$$

For a 99.9% confidence interval it is found as $\alpha = 1 - (99.9/100) = 0.001$.

Thirdly, use a statistical table or equivalent to find the position under the standard normal curve that has an area to the left of it equal to half alpha. For a 99.9% confidence interval this is the position that has $p = 0.001/2 = 0.0005$ of the curve to its left. We know, from Chapter 3, that distances from the centre of a standard normal are measured in z units so what we are looking for is the value of z that corresponds to $p = \alpha/2 = 0.0005$. Of the various candidates in Table 5.3, the most accurate is $z = -3.29$.

Fourthly, the z value is inserted into the equation defining the confidence interval:

$$P\left(\bar{x} - \left|z_{\alpha/2}\right| \times se_{\bar{x}} < \mu < \bar{x} + \left|z_{\alpha/2}\right| \times se_{\bar{x}}\right) = 1 - \alpha \tag{5.9}$$

Table 5.3 Part of a statistical table showing the area under a standard normal curve to selected values of z. High-lighted are *P*(*z* < −1.96) = 0.0250, *P*(*z* < −2.58) = 0.0049 and *P*(*z* < −3.29) = 0.0005, which are the number of standard errors needed around the sample mean for a 95%, 99% or 99.9% confidence interval respectively

	0.00	0.01	0.02	0.03	0.04	0.05	0.06	0.07	0.08	0.09
−3.2	0.0007	0.0007	0.0006	0.0006	0.0006	0.0006	0.0006	0.0005	0.0005	0.0005
...										
−2.5	0.0062	0.0060	0.0059	0.0057	0.0055	0.0054	0.0052	0.0051	0.0049	0.0048
...										
−1.9	0.0287	0.0281	0.0274	0.0268	0.0262	0.0256	0.0250	0.0244	0.0239	0.0233

This equation may seem intimidating but is only a more general expression of the formulae for the 95% and 99% confidence intervals that appear in Table 5.2. The $| \ldots |$ means that the negative sign can be ignored. For a 99.9% confidence interval for which $z = -3.29$ and $\alpha = 0.001$ the equation becomes

$$P(\bar{x} - 3.29se_{\bar{x}} < \mu < \bar{x} + 3.29se_{\bar{x}}) = 0.999 \tag{5.10}$$

This defines a 99.9% confidence. To apply it to a data set we need to know the mean and standard error. For the batch A data in Table 5.1 the sample mean is $\bar{x} = 1.106$ and the standard error is $se_{\bar{x}} = 1.226/\sqrt{36} = 0.204$. Rounding to two decimal places, the 99% confidence interval is

$$P(1.11 - 3.29 \times 0.20 < \mu < 1.11 + 3.29 \times 0.20) = 0.999$$
$$P(0.45 < \mu < 1.77) = 0.999 \tag{5.11}$$

Why is the alpha value divided by 2?

You may be wondering why the alpha value is divided by 2 in the third stage above, to find the value of z corresponding to $p = \alpha/2$.

The reason is that the population mean could be less *or* greater than the confidence interval predicts and therefore the probability of 'getting it wrong' – the alpha value – is split equally between the two tails of the normal curve. For example, a 95% confidence interval is based on the knowledge that $p = 0.95$ of the area under the curve is within $z = 2.58$ units of the centre. That leaves $p = 0.05$ of the area further away, of which half ($\alpha/2$) is found to the left of $z = -2.58$, and half (also $\alpha/2$) is found to the right of $z = +2.58$. This is illustrated by Figure 5.5.

5.11 Calculating a confidence interval for a small sample

Underpinning the definition of the confidence intervals is the expectation that the distribution of the sample means is normal. That expectation is founded on the central limit theorem but the theorem only applies if the sample size is not too small. When the size is less than about 30 ($n < 30$) then it is no longer reasonable to assume the distribution is normal. Instead, the sample means will follow a **t-distribution**.

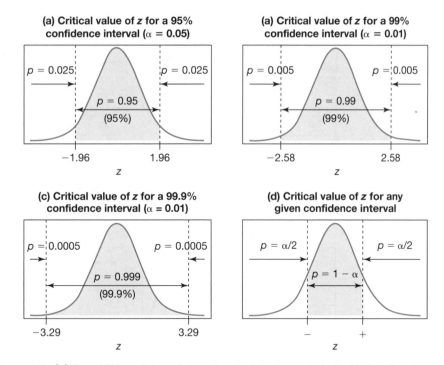

Figure 5.5 (a) For a 95% confidence interval we look in the statistical table for the value of **z** that has **p** = 0.025 to the left of it. (b) For a 99% confidence interval we look for the value that has **p** = 0.005 to the left of it. (c) For a 99.9% confidence interval we look for p = 0.0005. (d) More generally we look for the value of z corresponding to $p = \alpha/2$

The t-distribution was discovered by William Gosset and looks very similar to the normal distribution. However, it is 'fatter'. Quite how much so depends on the sample size or, more correctly, the degrees of freedom (Key concept 2.9). Here the degrees of freedom are simply one less than the sample size: $df = n - 1$. As the degrees of freedom decrease, the t-distribution spreads out more.

Figure 5.6 compares the shape of a t-distribution with that of a normal distribution. It was shown in the preceding discussion (and specifically Figure 5.5) that $p = 0.025$ of the area under a normal curve is to the left of $z = -1.96$. The corresponding value for the t-distribution is further from the centre. For a t-distribution with 10 degrees of freedom the value is -2.23 (to two decimal places), moving inwards to -2.06 when the degrees of freedom are raised to 25.

Similarly, we know that $p = 0.005$ of the area under a normal curve is to the left of $z = -2.58$. The corresponding value for a t-distribution with 10 degrees of freedom is -3.17, changing to -2.79 when the degrees of freedom are 25.

Figure 5.6 shows it is necessary to go further from the centre to fill 95% of the area of a t-distribution than for a normal distribution. The same is true to fill 99%, 99.9% or any other value we care to mention. The consequence of this is that confidence intervals are wider for small samples than for larger ones. Intuitively this makes sense: when we have a small amount of data we make a less precise estimate of the unknown population mean than we do when the data are more plentiful. That reflects the lesser amount of information available to support the inference.

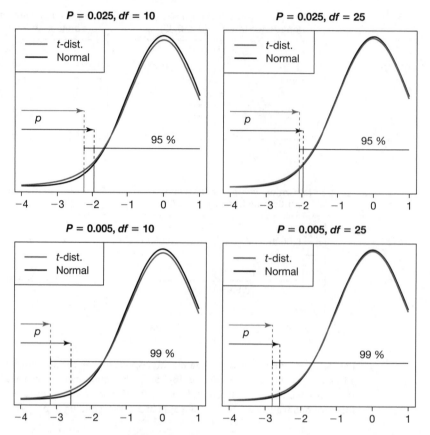

Figure 5.6 Comparing the area under the normal and *t*-distributions from the far left to the values of **p** associated with a 95% and 99% confidence interval. The *t*-distribution is 'fatter' than the normal, meaning that a greater proportion of its area is found further from the centre. Because of this, confidence intervals for small samples are wider than for larger samples. How much wider depends on the degrees of freedom

To reiterate, how much wider the confidence interval will be depends on the degrees of freedom. For a 95% confidence interval with 10 degrees of freedom, the interval needs to extend to 2.23 standard errors either side of the sample mean:

$$P(\bar{x} - 2.23se_{\bar{x}} < \mu < \bar{x} + 2.23se_{\bar{x}}) = 0.95 \tag{5.12}$$

For the same confidence interval but with 25 degrees of freedom, it needs to extend to 2.06 standard errors:

$$P(\bar{x} - 2.06se_{\bar{x}} < \mu < \bar{x} + 2.06se_{\bar{x}}) = 0.95 \tag{5.13}$$

More generally the confidence interval may be expressed as

$$P\left(\bar{x} - \left|t_{(\alpha/2, n-1)}\right| \times se_{\bar{x}} < \mu < \bar{x} + \left|t_{(\alpha/2, n-1)}\right| \times se_{\bar{x}}\right) = 1 - \alpha \tag{5.14}$$

which is analogous to the formula for a larger sample (Equation 5.9) but makes clear the need to look up a *t* value in a statistical table, where that value depends on the degrees of freedom, $n - 1$.

Table 5.4   Part of a statistical table showing the values of *t* required to form a 95%, 99% or 99.9% confidence interval for the stated degrees of freedom. Highlighted is the critical value required for a 95% confidence interval when the degrees of freedom are equal to 10

Degrees of freedom, $n-1$	$p = 0.0250$ (95% CI)	$p = 0.0050$ (99% CI)	$p = 0.0005$ (99.9% CI)
1	−12.706	−63.657	−636.62
2	−4.3027	−9.9248	−31.599
3	−3.1824	−5.8409	−12.924
4	−2.7764	−4.6041	−8.6103
5	−2.5706	−4.0321	−6.8688
10	−2.2281	−3.1693	−4.5869
15	−2.1315	−2.9467	−4.0728
20	−2.0860	−2.8453	−3.8495
25	−2.0595	−2.7874	−3.7251
30	−2.0423	−2.7500	−3.6460
99	−1.9842	−2.6264	−3.3915

A part of such a table is given by Table 5.4. It shows, for example, that the critical value for a 95% confidence interval and 10 degrees of freedom is $t = -2.2281$(to four decimal places).

Worked example

College Road is a street in the English town of Sandhurst and gets its name from the nearby Military Academy. According to house price data available at www.ourproperty. co.uk and using information supplied by the UK Land Registry, there were 21 properties sold in the street in 2007. The mean selling price was £276 167 and the standard deviation was £76 687. Based on these values, assuming they are representative and using a 99% confidence interval, what do we estimate the mean value of all properties in the street to be?

The sample size is $n = 21$. This is less than 30 so a *t* table is required. Using Table 5.4, we look for the value of *t* for $p = 0.005$ given 20 degrees of freedom ($= n - 1$). This is $t = -2.85$ (to two decimal places). The negative sign can be ignored.

The general expression for the confidence interval was given by Equation 5.14. We now know that $|t_{0.005}| = 2.85$, already knew that $\bar{x} = 276167$ and $s = 76687$, and can calculate the standard error as $s/\sqrt{n} = 76687/\sqrt{21} = 16734$. Substituting the values into Equation 5.14 gives the 99% confidence interval:

$$P(276167 - 2.85 \times 16734 < \mu < 276\ 167 + 2.85 \times 16734) = 0.99$$
$$P(276167 - 47692 < \mu < 276\ 167 + 47692) = 0.99$$
$$P(228475 < \mu < 323859) = 0.99 \qquad (5.15)$$

We estimate the mean property price for the street to be between £228 475 and £323 859 at a 99% confidence.

> ### Key concept 5.5 The *t*-distribution and confidence intervals for 'small samples'
>
> As a rule of thumb, a small sample is one that has less than 30 observations. For such samples we look up *t* values and not *z* values in a statistical table to form the confidence interval.
>
> The *t*-distribution and the normal distribution look similar but the *t*-distribution is 'fatter'. How much so depends on the degrees of freedom, $n - 1$.
>
> Use of the *t*-distribution widens the confidence interval around the sample mean of small samples. This reflects the increased uncertainty in the data relative to a larger sample.

5.12 Conclusion

In this chapter we have looked at the distinction between a sample of data and the population from which the sample is drawn. Whereas properties of the sample – the mean and the standard deviation – can be calculated in the usual way, the same properties for the population remain unknown unless we can survey it in its entirety.

We therefore make inferences about the population based on what we know about the sample. In particular, we can form confidence intervals and determine the probability that the population mean lies within a particular range of values. A logical next step is to ask, "Given what we know about the sample, is it plausible that the population mean could be equal to a theoretical value"? Or, "Given what we know about two samples, is it likely they were drawn from the same population"? This is the basis for hypothesis testing and is what we look at in the next chapter.

Before doing so, it is worth noting that the methods of inference we have demonstrated in this chapter make use of (long-run) probability and make an assumption of independence seen in, for example, the explanation of the central limit theorem given in Chapter 3: '[essentially] anything that can be thought of as being made up as the sum of many small *independent* pieces is approximately normally distributed' (Grinstead and Snell 2006, p.345 emphasis added).

Now consider our worked example using house prices. Are the prices paid for properties really independent of each other? Could there be geographical reasons to suppose they are not: social, cultural or economic factors, or the geography of the built environment (Orford 1999)? Might spatial autocorrelation and 'the first law of geography' be relevant here, the idea that 'things' in close proximity tend to have similar attributes and characteristics (Key concept 1.5)?

The answer to all these questions is, we suggest, 'yes!' Nevertheless, we will hold to the assumption of independence for many of the chapters and statistical tests that follow. Though doing so is not entirely satisfactory from a geographical point of view, we do it because such tests provide the foundations for the more explicitly geographical methods of analysis that come later, in Chapters 8 and 9.

Key points

- A sample is a subset of the population. The population is more abstract and may be understood as the complete set of all possible measurements that might (hypothetically) be recorded.

- Inference is at the heart of statistics. The aim is to infer, from a sample, properties of the population, notably its mean.

- We do not determine an exact value for the population mean but specify a range of values in which it might fall. That range is known as a confidence interval; the probability that the confidence interval includes the population mean can be calculated.

- The reason we can calculate the probability for larger samples is because of the central limit theorem and the properties of the normal curve.

- The standard error is a measure of unreliability. It increases with the standard deviation of the data and decreases with the square root of the sample size.

- For a fixed level of confidence, the confidence interval increases with the standard error.

- By convention, 95% and 99% confidence intervals often are used but not exclusively so.

- Small samples are those with less than 30 observations. For such samples the t-distribution is used instead of a normal curve to define the confidence intervals.

- Inference is always susceptible to error. We aim for the effects of error to be neutral overall and for the sample to be representative of the population from which it is drawn. Good sampling design is important here.

References

Boland, P.J. (1984) A biographical glimpse of William Sealy Gosset. *American Statistician,* 38(3), 179–183.

Crawley, M.J. (2005) *Statistics: An Introduction Using R,* Chichester: Wiley.

Frank, H. and Althoen, S.C. (1994) *Statistics: Concepts and Applications Workbook,* Cambridge: Cambridge University Press.

Grinstead, C.M. and Snell, J.L. (2006) *Introduction to Probability,* Providence, RI: American Mathematical Society. Available at: http://math.dartmouth.edu/~prob/prob/prob.pdf, accessed 20 October 2010.

Orford, S. (1999) *Valuing the Built Environment: GIS and House Price Analysis,* Farnham: Ashgate.

Salsburg, D. (2002) *The Lady Tasting Tea: How Statistics Revolutionized Science in the Twentieth Century,* 2nd edn, New York: Owl Books.

Hypothesis testing

Chapter overview

In Chapter 5 we emphasised the difference between a sample and the population from which the sample is drawn. We noted that whilst the mean of the sample can be calculated from the data, the mean of the population usually is unknown.

In this chapter we introduce hypothesis testing as a way of asking whether the population mean is equal to some hypothesised value.

The basic idea is to look at the difference between the sample mean and the hypothetical value. When doing so, we need also to consider the variability of the data. If the population generates highly varied data it could lead to a greater difference between the sample and the true mean than would arise from a more stable population.

Finding that the sample mean differs from the hypothesised value is not, in itself, proof that the hypothesised value is wrong. Some difference is inevitable given the variability of the population and the use of sampling to measure it. What we therefore determine is the probability that the difference has arisen by chance. The probability is lower the greater the difference, the less varied the data are, and the more data we have to support our conclusions. If that probability falls below a certain threshold we say the difference is statistically significant.

The logic can be extended to testing two or more samples. If their means are very different given the variability of the data and the amount of data we have then it is unlikely they were drawn from the same population. The inference is that the samples are measurements of categorically different things.

If there is a difference between samples it could be due to the effect of some intervention – for example, administering a new drug to some patients whilst giving a placebo to others. An effect does not have to be large to be important, but smaller effects are harder to find. Attention must therefore be given to how the research is designed and, in particular, to ensuring sufficient data are collected to enable an effect to be detected. Users of large secondary data sets may need to consider the reverse problem – of having so much data that almost all their results appear statistically significant, even those that have no obvious real-world meaning.

Learning objectives

By the end of this chapter you will be able to:

- Appreciate that because the properties of a sample depend on when, where and how it was taken, differences between samples will arise by chance even when they are measuring the same population.

- Understand that what matters is the magnitude of the difference, relative to the variability of the data and the amount of data collected. A small difference is hardly surprising if the population generates high levels of variation in the data. On the other hand, small differences can be significant if they exceed what we expect of the population.

- Describe the stages required to complete a formal process of hypothesis testing, determining whether a difference between one sample mean and a hypothesised value, or between two or more sample means, is sufficiently unlikely to be due to chance that it can be adjudged significant.

- Know the difference between a two-tailed and one-tailed test.

- Explain what is meant by Type I and Type II errors, know how they relate to statistical power, and be aware that conventional hypothesis testing can produce conclusions that are almost predetermined by too small or too large samples.

- Understand how an analysis of variance (ANOVA) is used to examine if three or more groups have the same population mean.

- Use an F test to compare variances.

- Know when to use non-parametric tests, for highly skewed or non-continuous data, for example.

- Recognise that the equations presented in this chapter repeatedly use information about the mean, standard deviation and standard error of one or more samples, these being the expressions of central tendency, spread and unreliability we discussed in previous chapters.

6.1 Detecting difference

We begin, as in other chapters, with some data. They are a random sample of all low-attaining pupils attending any one of the state-funded elementary schools within a particular metropolitan area during the period 2002–2006.

By low attainment we mean the pupils scored in the lowest decile (the bottom 10%) in a combined measure of results achieved in standardised tests of maths, English and science taken by the children across the city in their final year of elementary school. The sample is of length $n = 100$, which is about 6% of all low-attaining pupils. The data have been supplied by the Department for Education in England and Wales.

For each pupil a measure of income deprivation within their local neighbourhood also has been calculated. This is the square root of the proportion of children in families that are classified as income deprived. The square root is taken because, without the transformation, the percentages exhibit a mild positive skew (see Chapter 3, Section 3.9).

Figure 6.1 shows the distribution of the deprivation scores for the sample of pupils. It is approximately normal. In Figure 6.1(d) the mean is shown, together with the 95% and 99% confidence intervals (see Chapter 5). Finally, the dotted vertical line indicates the mean income deprivation score for all neighbourhoods within the city. This is an imagined value for the purpose of discussion though it is not dissimilar to the actual value for the study region.

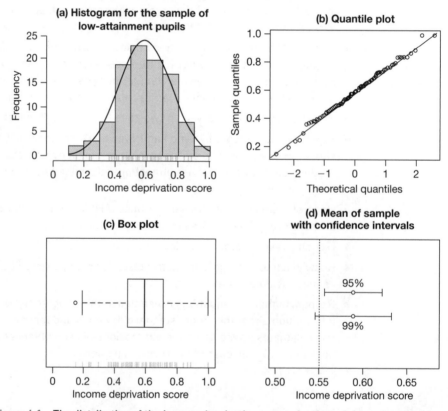

Figure 6.1 The distribution of the income deprivation scores for the neighbourhoods of a sample of low-attaining elementary school pupils. In (d) confidence intervals are shown around the sample mean, and compared with a hypothesised score of 0.55 for all neighbour-hoods in the study region

Source: Authors' own calculations based on data supplied by the Department for Education.

Still looking at Figure 6.1(d), there is a difference between the sample mean and the mean income deprivation score for the entire city. This may imply that low-attaining pupils tend to be located in neighbourhoods that are characterised by different amounts of income deprivation than is average for the region. If so, it raises social and public policy questions about the educational disadvantage experienced by some pupils in some places and how those might be remedied. However, the difference may also be due to chance, a peculiarity of the sample that would disappear were we to draw another.

We cannot entirely discount the possibility that the difference arose due to chance, but we can determine the probability it did so. Assuming the mean income score for the low-attaining pupils really is the same as for the city, and the differences have arisen owing to random sampling, we determine how frequently the difference observed would be exceeded in the long run, were we able to sample repeatedly from the population. If it is very often, the difference is not unusual and not statistically significant. If the difference rarely is exceeded then it can be regarded as **statistically significant**.

The clue to how to go about this is given by the confidence intervals shown around the sample mean. The 95% confidence interval does not include the mean neighbourhood score of 0.55. This suggests we can be 95% confident the sample is not from a population with mean $\mu = 0.55$. In other words, the probability that the difference is due to chance is $p \leq 0.05$ (5% or less).

The 99% confidence interval, however, *does* include the mean neighbourhood score. Therefore the probability that the difference arose by chance cannot be as low as $p = 0.01$ (1%). It is between the two.

In both cases, the probability is determined from the chosen level of confidence and from what we know about the sample: its mean ($\bar{x} = 0.59$), standard deviation ($s = 0.17$) and length ($n = 100$). That information is used to calculate the standard error ($se_{\bar{x}} = s/\sqrt{n}$) and to derive the confidence interval, as discussed in Chapter 5.

6.2 Hypothesis testing and the one-sample *t* test

The null and alternative hypotheses

Hypothesis testing is a process of structured enquiry that rejects or retains a clearly defined statement, using probability to guide the decision. It begins with a **null hypothesis**, a statement that (usually) we hope to disprove. This is a statement of no difference, of things being equal. Taking the hypothesis 'the mean income deprivation score is 0.55 for the sample of low-attaining pupils', it is written in statistical shorthand as

$$H_0 : \mu = 0.55 \qquad (6.1)$$

where H_0 denotes the null hypothesis and μ is the hypothesised value for the population mean, the hypothesised true mean income score for all low-attaining pupils in the study region.

Next, an **alternative hypothesis** is stated. This is the opposite of the null hypothesis – that the sample is from a population with mean attainment *not* equal to 0.55. Denoting the alternative hypothesis as H_1, it may be written as

$$H_1 : \mu \neq 0.55 \qquad (6.2)$$

The null hypothesis is rejected and the alternative hypothesis accepted only if there is sufficient evidence to do so. That evidence is generated by undertaking a statistical test and asking, 'if the population mean is equal to the hypothesised value, then what is the (long-run) probability of achieving a test result more extreme than the one we will calculate?'

If the probability is low, it suggests we have a rare result and therefore it is unlikely the population mean is equal to the hypothesised value. The null hypothesis can be rejected. The evidence is not definite proof, but it does allay reasonable doubt of making the wrong decision.

Logically the process of rejecting H_0 and accepting H_1 works only if the one statement being false necessitates the other to be true. Care is required to ensure we do not reason further than the logic allows. For example, though we may disprove the statement 'all swans are black' by observing a white swan, failing to observe a white swan does not prove they are all black.

The ornithological abstraction (much liked by philosophers) applies to how we interpret a failure to reject the null hypothesis. Not having the evidence to reject the null hypothesis does not mean that the null hypothesis is actually correct. It means *only* that there is insufficient evidence to reject it. The process is analogous to the principle of innocent until proven guilty used in most courts of law. We reject the null hypothesis only when the weight of evidence is beyond reasonable doubt. However, a problem arises if we make passing the test so stringent it becomes almost impossible to generate the evidence. This is a problem we return to later when we discuss statistical power.

The alpha value

Specifically, the null hypothesis is rejected when the probability of obtaining a test result more extreme than the one observed is less than a chosen alpha value. What we try to avoid is a **Type I error**. This is the act of *incorrectly* rejecting the null hypothesis when it is, in fact, true.

Key concept 6.1 Type I errors

The idea behind hypothesis testing is that the null hypothesis only is rejected if the balance of probability firmly is against its being correct. Nevertheless, there remains the possibility that rejecting the null hypothesis was the wrong decision. That mistake is known as a **Type I error**.

It cannot be known for certain whether a Type I error has been committed because we do not have complete information about the population. However, selecting an alpha value that gives high confidence in the test statistic not arising by chance reduces the prospect of committing a Type I error.

The alpha value is the maximum probability of committing a Type I error that we are prepared to tolerate and it defines the confidence level. It is the proportion of times we would incorrectly reject the null hypothesis given an infinite number of tests. Where possible, it is chosen at the outset, at a level appropriate to the study. In practice, two conventions dominate the literature: $\alpha = 0.05$ (a 95% confidence) and $\alpha = 0.01$ (a 99% confidence).

The test and the test statistic

Having decided on H_0, H_1 and α, a test is required. Here it will be a **one-sample *t* test**, measuring the number of standard errors the sample mean is away from the hypothesised value. The further it is, the less probably the sample is from a population with mean equal to the hypothesised value. The calculation is

$$t = \frac{\bar{x} - \mu}{se_{\bar{x}}} \tag{6.3}$$

where

$$se_{\bar{x}} = \frac{s}{l} \tag{5.3}$$

Testing for the null hypothesis of H_0: $\mu = 0.55$, Equation 6.3 becomes

$$t = \frac{0.59 - 0.55}{0.17 / \sqrt{100}}$$

$$= \frac{0.04}{0.017} = 2.35 \tag{6.4}$$

The result is called the **test statistic**. It is positive because the low-attaining pupils live in neighbourhoods that have, on average, a higher income deprivation score than the hypothesised value. However, that could be a chance characteristic of the sample. Therefore, the next step is to consider how frequently the magnitude of the test statistic would be exceeded (by chance) if we could repeatedly sample from a population with mean equal to the hypothesised value. If it is infrequently exceeded then the test result is unusual and unlikely due to chance.

The magnitude of the test statistic is its value ignoring any negative sign. It is indicated by the symbols $|...|$. More obviously, a value of $t > 2.35$ is greater in magnitude than $|t| = 2.35$. However, so too is a value of $t < -2.35$. Consider, for example, $t = -3$. Ignoring the negative sign, the magnitude is $|t| = 3$, greater than $|t| = 2.35$.

Therefore, what we are calculating is the probability of obtaining a test statistic of value greater than 2.35 *or* less than -2.35. Using notation, we are determining

$$P(|t| > 2.35) = P(t > 2.35) + P(t < -2.35) \tag{6.5}$$

The reason for considering both negative and positive values is due to the alternative hypothesis. It states only that the sample is from a population with mean not equal to the hypothesised value. Whilst we may suppose the lower attaining pupils are from neighbourhoods with higher income deprivation on average, the test makes no such presupposition. It says the population mean could be greater or less than the hypothesised value. To test specifically that it is greater requires a slightly different approach (see Section 6.3).

Key concept 6.2 The one-sample *t* test

The **one-sample *t* test** calculates the difference between a sample mean and a hypothesised value and then considers the probability that the difference arose by chance. The greater the difference, the less varied the data are and the greater the number of observations, the lower the probability.

The critical value

It is not necessary to determine the exact probability of exceeding the test statistic. We need only know if it is less than the alpha value. If it is, the null hypothesis can be rejected because the probability of committing a Type I error is within what we are prepared to tolerate.

Setting the alpha value to be $\alpha = 0.05$, the critical value for our one-sample *t* test is $|t| = 1.98$. This is the distance we need to go out from the centre of a *t*-distribution, with 99 degrees of freedom, to be left with $p = 0.05$ of its area remaining, split equally between the two tails. You can confirm this value is correct by looking back at Table 5.4.

The critical value arises from the test and from the knowledge that if we could sample repeatedly from the population and calculate the test statistic each time, the distribution of the results would follow a *t*-distribution with $n - 1$ degrees of freedom. In fact, given the sample size and also the central limit theorem, we might reasonably assume it follows a normal distribution. If so, the critical value when $\alpha = 0.05$ will be $|z| = 1.96$ (we know this from Chapter 3, Figure 3.6, for example).

The reason for preferring the *t*-distribution (and the one-sample *t* test) over the normal distribution (and the one-sample *z* test) is because of how the standard error is estimated, using the standard deviation of the sample as a substitute for the unknown standard deviation of the population (see Chapter 5, Section 5.7). There is also a more pragmatic reason: the *t*-distribution is better for smaller samples, and for larger samples it and the normal curve are extremely similar anyway.

Rejecting or retaining the null hypothesis

If the test statistic has greater magnitude than the critical value it is the more extreme of the two and the less likely to be exceeded by chance. Given that the critical value is fixed at the maximum probability we are willing to accept for a result due to chance, finding the test statistic to be less probable provides the evidence needed to reject the null hypothesis. Written more simply, **if the test statistic exceeds the critical value, the null hypothesis is rejected.**

It is only rejected at the chosen level of confidence. For the pupil data, an alpha value of $\alpha = 0.05$ leads to a critical value of $|t| = 1.98$, exceeded by the test statistic, $|t| = 2.35$, and the null hypothesis is rejected. We conclude that the low-attaining pupils are living in neighbourhoods that have, on average, levels of income deprivation that are significantly different from the hypothesised value.

Changing the alpha value to $\alpha = 0.01$ gives a new critical value of $|t| = 2.63$, reflecting the increased confidence sought. In this case, it is no longer exceeded by the test statistic and the null hypothesis is not rejected.

Reporting a *p* value

The approach outlined above is the traditional one that compares the test statistic with a critical value and rejects or retains the null hypothesis accordingly. What this approach does not do is calculate a specific probability associated with the test statistic. It just determines whether it is above or below the alpha value. Working in this way avoids the need to publish statistical tables for every conceivable test statistic and degree is of freedom. However, it also loses information in the sense that a computer can readily calculate an exact probability, which is called the **p value**.

The probability of drawing a value of $|t| > 2.35$ from a t-distribution with 99 degrees of freedom is $p = 0.021$ (2.1%). This is the probability of committing a Type I error.

There is a good argument for reporting the *p* value and not just stating if the null hypothesis is rejected. Doing so allows one to work backwards and determine the maximum level of confidence that can be attached to rejecting the null hypothesis. For our example it is 97.9% ($= 100\% - 2.1\%$).

Being open about this information prevents us committing a statistical sleight of hand, selecting the 95% confidence interval just because it rejects the null hypothesis (and concealing that the 99% confidence does not). Even so, there is still advantage in fixing the alpha value in advance of the test – it encourages us to be informed by previous research about the level of confidence appropriate for the study.

The process of hypothesis testing is summarised in Key concept 6.3.

Key concept 6.3 Hypothesis testing

Hypothesis testing is a formal process of asking whether a logical statement called the **null hypothesis** should be rejected in favour of an opposite statement, the **alternative hypothesis**.

The stages of hypothesis testing are to:

1. Define the null hypothesis, H_0.
2. Define the alternative hypothesis, H_1 (the logical opposite of H_0).
3. Specify an **alpha value**, α, which is the maximum probability we are willing to accept of committing a Type I error.
4. Calculate the **test statistic**.
5. Compare the test statistic with a **critical value**.
6. Reject the null hypothesis if the test statistic is of greater magnitude than the critical value.

In addition to stages 4 and 5, it is useful to report a **p value** giving an exact estimate of the probability that the test statistic has arisen by chance.

6.3 One-tailed tests of difference

The one-sample *t* test described above asked only if the sample of low-attaining pupils is from neighbourhoods with a mean income poverty score different from the hypothesised value. It did not raise any expectation about whether the mean would be higher or lower. The test was non-directional.

Yet, our expectation is that there is a relationship between income poverty and educational disadvantage, and that low-attaining pupils are found in neighbourhoods with higher income poverty, on average. Therefore, a better test would be directional – looking for evidence that the sample is from a population with mean *greater* than the hypothesised value. The null hypothesis is as before (H$_0$: $\mu = 0.55$) but the alternative hypothesis changes:

$$\text{H}_1 : \mu > 0.55 \tag{6.6}$$

(cf. Equation 6.2).

The calculation of the test statistic is the same, still using Equation 6.3 and therefore still obtaining $t = 2.35$ with 99 degrees of freedom. However, the critical value has changed. Previously it was $t = 1.98$ for $\alpha = 0.05$. Now it is $t = 1.66$. We need to understand why.

Figure 6.2 offers an explanation. The left-hand graphs show how the critical value is determined for a 95% confidence interval when the alternative hypothesis is non-directional (when we test for a difference but have no initial expectation about whether the hypothesised value will be greater or less than the sample mean). For such tests the maximum probability of committing a Type I error, the alpha value, is split equally between *both* tails of the *t*-distribution. This is why the non-directional test is said to be two tailed.

In contrast, the right-hand graphs show the critical value for a directional test, still with a 95% confidence but with the alternative hypothesis stating that the sample is from a population with mean *greater* than the hypothesised value.

The directional test is one tailed. Although the probability of committing a Type I error has not been changed, it is now contained solely within one tail. To accommodate this, the critical value moves inwards. The principle remains the same when the alternative hypothesis states that the sample mean is less than the hypothesised value, except then the probability of committing a Type I error moves to the other tail.

There is another way of distinguishing a non-directional test from a directional one. The first works by creating a confidence interval with the sample mean at the centre. We then determine whether the hypothesised value falls within the confidence interval and, if it does not, reject the null hypothesis. The 95% confidence interval for the income deprivation scores of the low-attaining pupils is $P(0.556 < \mu < 0.622) = 0.95$ and is shown in Figure 6.1(d).

The directional test is similar, except because it is one tailed the confidence interval is no longer centred on the sample mean. Instead, it is shifted sideways to begin or end at infinity. For the sample of low-attaining pupils it becomes $P(0.561 < \mu < \infty) = 0.95$.

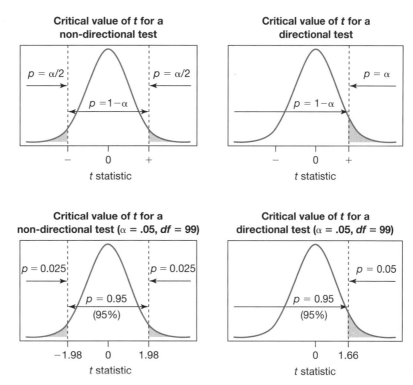

Figure 6.2 Illustrating why the critical value is not the same for a directional test as it is for a non-directional test. The non-directional test is two tailed: the probability that the result is due to chance is split equally between the tails. For a directional test, that probability moves solely into one tail

---

Key concept 6.4 Two- and one-tailed tests

A **two-tailed test** is **non-directional** whereas a **one-tailed** test is **directional**.

For a one-sample t test the null hypothesis is that the sample is drawn from a population with mean equal to the hypothesised value. For a **two-tailed** test the alternative hypothesis is that they are not equal, but which is the greater is not specified. For a **one-tailed** test the alternative hypothesis says that one is greater (or less) than the other.

Whether you should use a one- or two-tailed test depends on your research question. For example, expecting girls to do better than boys, on average, in high school examinations in the UK suggests a directional test. Expecting a difference but not knowing which gender does better requires a two-tailed test.

6.4 ## Power, error and research design

In our description of hypothesis testing, emphasis is given to the alpha value, α, as the basis for rejecting the null hypothesis (or not). It sets the probability of rejecting the null hypothesis when it is, in fact, true – of committing a Type I error.

On face value, it is a sensible strategy. Consider a medical trial where one group of patients is given a new drug and a second receives a placebo. Before the start-up costs of mass production and marketing, and before raising the hopes of prospective users, the pharmaceutical company needs to be sure the drug is effective by making comparative measurements of the two groups of patients and looking for a statistically significant difference, one that is unlikely due to chance alone. The difference is **the effect** of the drug. By limiting the possibility of a Type I error no claim is made for the effectiveness of the treatment until the balance of the evidence is firmly against the observed effect being a fluke.

Unfortunately, there is a problem. Think for a moment about how you could ensure a Type I error will not occur. How would you go about it? If it helps, return to the court analogy and consider how a jury could ensure it would never wrongly declare someone guilty. Think it over before you read on.

The answer is simple: never reject the null hypothesis. Just do not do it. Ever. It is, of course, ludicrous. The null hypothesis could be wrong and deciding otherwise from the outset renders the hypothesis test purposeless. A system that always acquits the defendant is not justice – it may protect the innocent but it also always sets the guilty free.

Here is the point: though it is a mistake to reject the null hypothesis when it is correct (a Type I error), it is also a mistake to retain the null hypothesis when it is wrong. The latter is a called a **Type II error**. If we become overly cautious about rejecting the null hypothesis we also make it hard to accept the alternative (even when the null hypothesis is wrong).

A Type II error can also be understood as the act of rejecting the alternative hypothesis when it is correct. If the probability of committing a Type II error is given the symbol β then that leaves $(1 - \beta)$ as the probability of making the correct decision – of accepting the alternative hypothesis when it is, indeed, correct.

The distinction between Type I and Type II errors is shown in Table 6.1.

In statistical language, $(1 - \beta)$ defines the **power** of the test. A power level of 0.8 is usually regarded as desirable and indicates an incorrect null hypothesis will rightly be rejected four times out of five.

Returning to the medical trial, power is the probability of detecting an effect that would be true in the population as it appears in the sample. In short, power is

Table 6.1   The possible outcomes of a statistical test. In practice we do not know for certain whether the null hypothesis is true or not (if we did there would be no need for the statistical test!)

		Should the null hypothesis have been rejected?	
		No	Yes
Is the null hypothesis rejected?	No	Correct decision	Incorrect decision: a Type II error $P(Type\ II\ error) = \beta$
	Yes	Incorrect decision: a Type I error $P(Type\ I\ error) = \alpha$	Correct decision power $= (1 - \beta)$

the probability of detecting something genuine. As Murphy and Myors (2003, p.6) observe:

> the power of a statistical test is a function of its sensitivity, the size of the effect in the population, and the standards or criteria used to test statistical hypotheses.

All else held constant, if you minimise Type I errors by decreasing the alpha value then you also reduce the power of the test. Whilst an argument can be made for caution, of not rushing to reject the null hypothesis, to do so is to assume the principal cost of a decision is in the acceptance of the alternative hypothesis – that there is little to lose by sticking with the null hypothesis. In practice, that is unlikely to be the case.

For a business there are issues of innovation and competitive advantage to consider. As a medical patient you might regard any possibility of a positive effect as better than none at all, and any level of confidence over 50% places the balance in your favour. Murphy and Myors (2003) argue that the conventions of $\alpha = 0.05$ and $\alpha = 0.01$ are often too strict and ought to be relaxed (to $\alpha = 0.10$ for example).

Furthermore, the rejection of the null hypothesis is a function of the sample size. This can be demonstrated with a simple simulation.

Consider a random sample of $n = 50$ observations drawn from a population with mean $\mu_A = 25.00$ and standard deviation $\sigma = 1$. Imagine a t test is used to compare the sample with a second, also of $n = 50$ observations but this time drawn from a population with mean $\mu_B = 24.99$ (and still $\sigma = 1$). We introduce the two-sample t test fully in Section 6.5. Here it is sufficient to know it is used to determine whether two samples could have the same population mean. The null hypothesis is that they do: $H_0: \mu_A = \mu_B$.

Now imagine the process of sampling and testing is repeated to give a total of 100 separate tests, each with samples of size $n = 50$. We then count up how many times of the hundred the null hypothesis is rejected, using an alpha value of $\alpha = 0.05$.

The result of the simulation is shown in the first row of Table 6.2. The null hypothesis was rejected seven times. That seems about right given the random sampling, that the population means are only 0.01 units apart and given a 95% confidence implies that differences generated by chance will test as significant about five times in every hundred.

Table 6.2 Simulating the effect of sample size on the power of a two-sample t test, where each random sample is of size n, the null hypothesis is $\mu_A = \mu_B$, and where $\mu_A = 25.00$ and $\mu_B = 24.99$. As the sample size increases, so too does the power of the test and the null hypothesis more frequently is rejected

n	Number of times, from 100 independent tests, H_0 is rejected ($H_0: \mu_1 = \mu_2$)
50	7
500	4
5000	9
50 000	39
500 000	100

Similarly, when we repeat the entire simulation but now using sample sizes of $n = 500$ and then $n = 5000$, we find the null hypothesis is rejected four times and nine times respectively. Those results are shown in the second and third rows of Table 6.2. However, when we raise the sample size to $n = 50\,000$ the null hypothesis is rejected 39 times. When $n = 500\,000$ it is rejected *every* time. What is happening?

The fact is that there *is* a difference between the population means. It is slight at 0.01 units but it exists. The statistical test incorporates a measure of unreliability, the standard error, which is determined, in part, on the size of the samples.

As Murphy and Myors (2003, p.16) note:

> [I]t is tempting to conclude that a significance test is little more than a roundabout measure of how large the sample is. If the sample is sufficiently small, the null hypothesis would virtually never be rejected. If the sample is sufficiently large, the null hypothesis will virtually always be rejected.

What they are saying is that the power of the statistical test increases with the sample size. The larger the sample size, the more often we expect the null hypothesis to be rejected. This immediately raises questions about research design (see Chapter 4) and avoiding research that is doomed to failure at the outset. If the sample sizes are too small to detect subtle differences or hypothesised effects then failure to reject the null hypothesis should come as no surprise: it was essentially predetermined before the analysis even began!

For example, in an influential paper within geomorphology, Strahler (1952) uses *t* tests to compare the topography of river basins in various American localities based on measurements of what is called the hypsometric integral. In each locality six basins are sampled and the samples then tested two at a time to look for differences between localities, differences that are then related to geomorphic processes.

If we take the difference between Verdugo Hills, CA and Great Smoky, NC, the reported *p* value of 0.08 would not conventionally be considered significant. However, with a sample size of six and a value of $\alpha = 0.05$ the power of the test is less than 0.5: it is more likely the null hypothesis will not be rejected even where a difference between the localities exists. Bringing the power of the test to the preferred level of about 0.80 requires either an alpha value of about $\alpha = 0.25$ or a sample size of about 15.

The key point here is that research needs to be designed so sufficient data are collected that any effect that might exist can be detected. It is not necessary to know exactly what the size of the effect will be in advance of the study, only to have an informed estimate, based on previous studies. This permits the sample size, the alpha value and the power of the test to be investigated at the planning stage, using an online tool such as the one by Lenth (2006–2009) at http://www.stat.uiowa.edu/~rlenth/Power. Doing so helps prevent too little data from being collected.

Reciprocally, users of the large social and environmental data sets (many available to download from online data depositories) may encounter a problem of 'too much power'. Analysts of such data should not be surprised, for example, if a regression analysis (see Chapter 7) reveals many significant effects. Statisticians who devised tests and conventions during a period when 'large sample' could mean more than 30 observations could not have anticipated their use on secondary data containing thousands or millions of observations. That we now do so means we risk attributing significance to results that have no real or substantive meaning.

The problem of too much power arises because the null hypothesis is also a 'nil hypothesis' (Cohen 1988). It is a definition of no effect, of no difference between the population means. The trouble is that almost any medical intervention (for example) has some effect, albeit a placebo effect. Therefore, a better null hypothesis is not of *zero* effect but of *negligible* effect, where what is considered negligible depends on the context of the analysis. Testing in this way is beyond the scope of this book, but working through our earlier simulation and applying the methodology outlined by Murphy and Myors (2003) we find that even when $n = 500\,000$ we never reject a null hypothesis of negligible difference. The rejection of the null hypothesis is no longer predetermined by the sample size.

Key concept 6.5 Type II errors and statistical power

Accepting the null hypothesis when it is, in fact, wrong, is known as a **Type II** error.

The ideal, of course, is to reject the null hypothesis when it is wrong, and the probability of doing so is the **power** of the test.

When conventional alpha values are applied to relatively small data sets the power of the test is low, especially when the effect we are hoping to detect is itself small. This can lead to the erroneous conclusion that there is no effect, whereas the failure to reject the null hypothesis can actually be due to how we designed the research.

On the other hand, users of very large secondary data sets may find that almost everything they test is significant. This is a function of the sample size and our increased confidence to detect small differences. What the analyst needs then to consider is whether the effect is essentially negligible, and whether it has real substantive or theoretical meaning.

6.5 The two-sample *t* test and the *F* test

Figure 6.3 compares the centre and spread of two samples of data. The one labelled 'bottom decile' is the sample of Section 6.3 giving the neighbourhood income deprivation scores of pupils who scored in the bottom 10% of results from a standardised test of attainment administered in elementary schools. The sample labelled 'top decile' is of pupils who scored in the top 10%. Both samples happen to contain $n = 100$ observations but they need not be equal in number for the statistical test to proceed.

Looking at Figure 6.3 we find the higher attaining pupils are generally from less income-deprived neighbourhoods and vice versa. However, the box plots also show the samples overlap. There are variations between pupils. Finding this variation is not surprising – people vary in their academic ability – and it is not, itself, a reason to discount the general trend. What matters is the amount of difference between the sample means relative to the variation found within the data and the amount of data we have.

To assess this, the **two-sample *t* test** is used. The logic is the same as for the one-sample *t* test except the second sample replaces the hypothesised value. If the sample means are a long way apart relative to the variation in the data, and given the amount of data we have, then it is unlikely the samples belong to the same population. This would suggest that the difference is statistically significant and provides evidence that

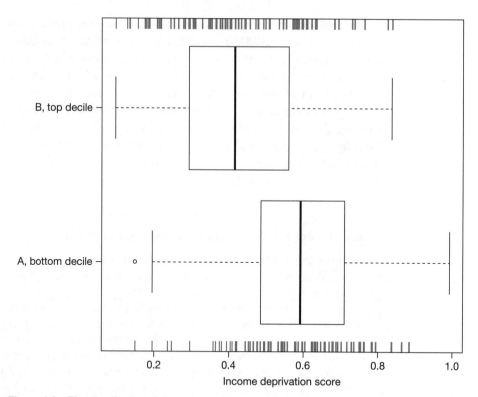

Figure 6.3 The distribution of the neighbourhood income deprivation scores for two samples of pupils. Sample A is drawn from the highest 10% of test results. Sample B is drawn from the lowest 10%

Source: Authors' own calculations based on data supplied by the Department for Education.

the samples have categorically different levels of income deprivation. The test is now developed in stages.

Stage 1: Forming the test

To begin, recall how the one-sample t test calculated the difference between the sample mean and the hypothesised value, relative to a measure of the data's unreliability, the standard error. We can summarise the test statistic as

$$t = \frac{\text{difference}}{\text{unreliability}} \tag{6.7}$$

For a two-sample t test, the difference is between the sample means:

$$\text{difference} = (\overline{x}_A - \overline{x}_B) \tag{6.8}$$

The measure of unreliability becomes **the standard error of the difference** (SED) and brings together information about each sample's length and standard deviation:

$$\text{unreliability (SED)} = \sqrt{\frac{s_A^2}{n_A} + \frac{s_B^2}{n_B}} \tag{6.9}$$

Substituting Equations (6.8) and (6.9) into (6.7) gives a test of the form

$$t = \frac{\overline{x}_A - \overline{x}_B}{\sqrt{\dfrac{s_A^2}{n_A} + \dfrac{s_B^2}{n_B}}} \qquad (6.10)$$

As with the one-sample *t* test, the value will less frequently be exceeded in the long run the greater the difference, the less the variation in the data, and the more observations there are.

Unfortunately, if the two variances (s_A^2 and s_B^2) are unequal then the result is not really a *t* statistic, though it does approximate to being so with degrees of freedom calculated using a formula outlined by Welch (1949). Alternatively, if s_A^2 and s_B^2 are equal, or approximately so, then it can be better to calculate the standard error of the difference in another way, using the pooled standard deviation, s_p, instead of the individual sample values:

$$\text{SED} = s_p \sqrt{\frac{1}{n_A} + \frac{1}{n_B}} \qquad (6.11)$$

Back in Chapter 2 we showed that the standard deviation for a (single) set of data is a measure of average deviation from the mean and is calculated as the square root of the sum of squares divided by the degrees of freedom (see Key concept 2.10). The pooled standard deviation is the same except the sum of squares and the degrees of freedom are for the two samples together,

$$s_p = \sqrt{\frac{\displaystyle\sum_{i=1}^{n_A}\left(x_i - \overline{x}_A\right)^2 + \sum_{i=1}^{n_B}\left(x_i - \overline{x}_B\right)^2}{\left(n_A - 1\right) + \left(n_B - 1\right)}} \qquad (6.12)$$

Using s_p then allows the *t* statistic to be determined as

$$t = \frac{\overline{x}_A - \overline{x}_B}{s_p \sqrt{\dfrac{1}{n_A} + \dfrac{1}{n_B}}} \qquad (6.13)$$

with $(n_A + n_B - 2)$ degrees of freedom. This can be more accurate than using Equation 6.10 because s_p is pooling information from both sets of data. However, it is doing so on the assumption that the populations have an equal standard deviation, an assumption that might not be valid.

In practice you are unlikely to need Equations 6.10 to 6.13 because the statistical package will do the calculations for you. You are still left with a decision: do you assume the populations have equal standard deviations – and therefore equal variances – or not?

Stage 2: Testing for equal variance (an *F* test)

An *F* test is used to determine if two samples are drawn from populations with approximately equal variance. Equal variance is known as **homoscedasticity**, and this is the null hypothesis for the test. The opposite, unequal variance, is called **hetero-scedasticity** and provides the alternative hypothesis. Thus,

$$H_0 : \sigma_A = \sigma_B$$
$$H_1 : \sigma_A \neq \sigma_B \tag{6.14}$$

The basis for the test is simple: divide one sample's variance by the other and see if the answer is one, in which case they must be equal. For consistency, divide the larger variance by the smaller. This ensures the test statistic has a value equal to or greater than one:

$$F = \frac{s_A^2}{s_B^2} \ \left(\text{if } s_A^2 > s_B^2 \right) \ \text{or} \ F = \frac{s_B^2}{s_A^2} \ \left(\text{if } s_A^2 < s_B^2 \right) \tag{6.15}$$

As ever, the result is dependent upon the samples. Finding that the result is not exactly equal to one does not prove the populations also have unequal variances; the inequality could be due to chance. We therefore do as we did for the t test and determine the long-run probability of obtaining a test statistic more extreme than the one we have. If the probability is low it provides evidence to reject the null hypothesis of equal population variance and to accept the alternative hypothesis instead.

The reference distribution for the probability calculations is the **F-distribution**. Examples of the F-distribution are shown in Figure 6.4. Note that it has a positive skew, that F values are always positive and that the exact shape of the curve is determined by two degrees-of-freedom values, df_1 and df_2, one for each of the variance values in Equation 6.15.

Comparing the variances for the two samples of pupils gives

$$F = \frac{s_B^2}{s_A^2} = \frac{0.030}{0.027} = 1.11 \tag{6.16}$$

The statistic has 99 degrees of freedom in the numerator (the top part of Equation 6.16) and 99 degrees of freedom in the denominator (the bottom part). The degrees of freedom are calculated as $n_B - 1$ and $n_A - 1$.

Critical values for the F statistic are awkward to show in a statistical table because it must allow for the two degrees-of-freedom values as well as the required level of confidence. The traditional solution is to present multiple tables, one each for specific alpha values, within which the critical value of F for selected degrees of freedom is stated. However, we can also use statistical software, in this case obtaining an exact value of $p = 0.663$.

The p value is too great to reject the null hypothesis of equal variance. We therefore proceed with a t test assuming equal variance of the populations.

Stage 3: Completing the t test

Recall that we are testing whether the mean level of income deprivation for the two samples of pupils is the same and have determined that a two-sample t test assuming

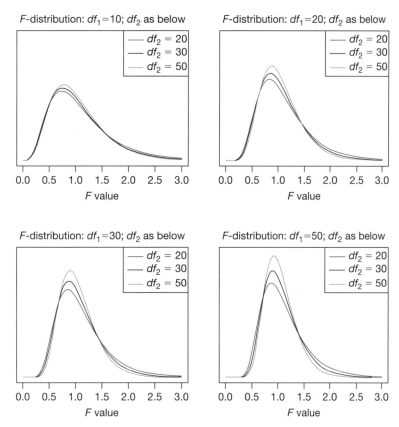

F-distribution: $df_1=10$; df_2 as below

F-distribution: $df_1=20$; df_2 as below

F-distribution: $df_1=30$; df_2 as below

F-distribution: $df_1=50$; df_2 as below

Figure 6.4 The shape of an *F*-distribution. It is related to both the normal and the *t*-distributions but has a positive skew, the extent of which is governed by the two degrees-of-freedom values

Note that the curve for $df_2 = 20$ is the bottom curve in each graph, $df_2 = 30$ is the middle one, $df_2 = 50$ is the top one.

Key concept 6.6 Homoscedasticity, heteroscedasticity and the *F* test

When data have equal variance they are said to be **homoscedastic**. When unequal, they are **heteroscedastic**. The sample variance (s^2) is the square of its standard deviation (s).

The simplest test of whether two samples have equal variance is to divide one sample's variance by the other and see if the result is equal to one. However, they are always unlikely to be exactly equal.

To determine whether they are significantly different, an **F test** is undertaken that compares the ratio of the sample variances with a critical value drawn from an *F*-distribution with $(n_A - 1)$ and $(n_B - 1)$ degrees of freedom. The test is used to reject or to retain the null hypothesis that the samples are from populations of equal variance.

---

equal variance is appropriate to use. Given our expectation that income poverty is related to educational disadvantage, we will undertake a one-tailed test to see if sample A (of the lower attaining pupils) has a higher level of income deprivation than sample B (of the higher attaining pupils). The null and alternative hypotheses, and the alpha value, are as follows:

$$H_0 : \mu_A = \mu_B$$
$$H_1 : \mu_A > \mu_B \qquad (6.17)$$
$$\alpha = 0.05$$

Using our statistical package we obtain a test statistic of $t = 6.99$ with 198 degrees of freedom, exceeding the critical value of $t = 1.65$. The null hypothesis is rejected. The sample of low-attaining pupils is drawn from more income-deprived neighbourhoods, on average. The probability we have made a mistake is, in fact, tiny: the p value is marginally above zero (but not zero; there is always a chance the result is due to chance).

6.6 Additional notes about the t test

A t test is used to test whether the mean of a normally distributed population is equal to a hypothesised value, or whether two normally distributed populations have an equal mean.

Seeing that the samples are normally distributed implies their populations will be too. We have used the quantile plot as a visual tool for checking normality and there are statistical tests that can be used too, including the Shapiro–Wilk and the Anderson–Darling test. In our particular software package (R) the null hypothesis for the tests is that the data *are* drawn from a normal distribution. It is not, therefore, a hypothesis we want to reject. Fortunately we do not have to. The p value for the Shapiro–Wilk test *exceeds $p = 0.05$* for both samples of pupils.

The two-sample t test assumes the samples are taken independently of each other. Whilst the assumption of *complete* independence is often suspect in geographical research because of the various social and physical processes that create the patterns we measure at various scales on the Earth's surface, still there will be occasions when dependence should be recognised explicitly; for example, when an observation in one sample directly is paired with an observation in a second sample.

Consider, for example, measures of species diversity (flora or fauna) taken at regular intervals on both sides of a field boundary, where on one side petrochemicals are used to manage crops and on the other more organic practices are trialled. Assume also that the field is on a slope, with the bottom more waterlogged and the higher more exposed to wind, and that some parts of the field receive sunlight whilst others are in the shade.

The hypothesis is that organic practices are associated with greater species diversity. However, it is anticipated that each pair of measurements will jointly be affected by the immediate environment and microclimate.

A **paired t test** calculates the difference in value between an observation and its pair, doing so for all pairs of data. The mean ($\bar{x}_D$) and standard deviation (s_D) of those differences is then calculated and a test statistic obtained against a null hypothesis of zero mean difference within each pair:

$$t = \frac{\bar{x}_D}{s_D / \sqrt{n}} \qquad (6.18)$$

This is a t statistic with $(n-1)$ degrees of freedom, where n is the number of pairs. In fact, the paired t test is simply a one-sample t test with a hypothesised value of zero (compare Equation 6.18 with 6.3).

Can you think of other situations when data are paired? How about a standardised test of learning administered to students in an elementary school and then, again, to the same students in high school?

6.7 Analysis of variance

An analysis of variance (ANOVA) is used to test the null hypothesis that three or more samples, groups or batches of data have an equal population mean:

$$H_0 : \mu_A = \mu_B = \mu_C = \cdots \tag{6.19}$$

To demonstrate the method, Figure 6.5 shows life expectancy, measured in years at birth, of males in 117 countries grouped by a measure of national wealth. The life expectancy data are accessible via the UN data portal (http://data.un.org) and are from Gender Info. 2007, a global database of gender statistics and indicators published by the United Nations Statistics Division. The country groupings are by the World Bank and are listed at http://go.worldbank.org/D7SN0B8YU0.

Looking at the data, it is clear that mean life expectancy is least in low-income countries, at $\bar{x}_A = 52.1$ years. That differs from lower middle-income countries for which the mean is $\bar{x}_A = 63.3$ years, and from upper middle-income countries where it is $\bar{x}_A = 67.3$ years. Yet, there is also variation around those means. The interquartile

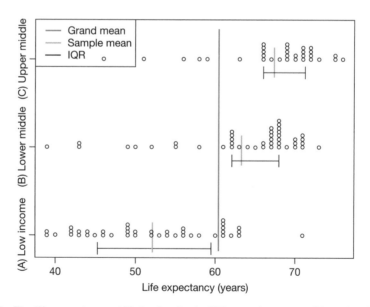

Figure 6.5 The life expectancy at birth of males in 117 countries grouped by a classification based on gross national income (GNI) per capita. The grand mean is the average across all data values regardless of their grouping

Data sources: UNdata and World Bank, 2009.

range (IQR) for the low-income countries is from 45 to 59.5 years, for example, and the IQRs for the lower and upper middle countries overlap.

An analysis of variance works by considering the variation *between* groups relative to the variation *within* groups. If the group means are all very similar and there is a lot of variation around those means then there is little evidence to suggest the groups are especially different. On the other hand, if the variation between groups is great and the variation within groups is low then the measurements will display little overlap, implying one or more of the groups is categorically different from the rest.

How, then, to quantify the variation within and between groups? For the variation within, our primary measure has been the sample standard deviation, which is based on the sum of squares, SS (see Key concept 2.8):

$$s = \sqrt{\frac{SS}{n-1}} \tag{6.20}$$

where

$$SS = \sum_{i=1}^{n}\left(x_i - \bar{x}\right)^2$$

Adding together the sum of squares for the three groups gives the total within groups sum of squares, abbreviated as WSS:

$$WSS = SS_A + SS_B + SS_C \tag{6.21}$$

Equation 6.21 can also be written as

$$WSS = s_A^2(n_A - 1) + s_B^2(n_B - 1) + s_C^2(n_C - 1) \tag{6.22}$$

because

$$SS = s^2(n-1)$$

(cf. Equation 6.20)

More generally, if there are k groups in total,

$$WSS = \sum_{j=1}^{k} s_j\left(n_j - 1\right) \tag{6.23}$$

The variation between groups is also measured using a sum of squares calculation. This is the sum of the squared difference between each group mean and the grand mean, allowing for group size. The grand mean is the mean for all observations taken together (regardless of their grouping) and is indicated by the long vertical line in Figure 6.5. Giving the grand mean the symbol $\bar{\bar{X}}$, the between-groups sum of squares (BSS) for our three groups is calculated as

$$BSS = n_A\left(\bar{x}_A - \bar{\bar{X}}\right)^2 + n_B\left(\bar{x}_B - \bar{\bar{X}}\right)^2 + n_C\left(\bar{x}_C - \bar{\bar{X}}\right)^2 \tag{6.24}$$

or, more generally, as

$$BSS = \sum_{j=1}^{k} n_j\left(\bar{x}_j - \bar{\bar{X}}\right)^2 \tag{6.25}$$

Finally, we determine the ratio of BSS to WSS, taking into account the degrees of freedom used to calculate each. This gives an F statistic with $(k-1)$ degrees of freedom

Table 6.3 Calculating the F statistic for an ANOVA of the life expectancy data. See text for details

	Low income	Lower middle income	Upper middle income	
n	42	43	32	$N = \Sigma n = 117$
$\bar{x}$	52.1	63.3	67.3	$\bar{\bar{X}} = 60.4$
$(\bar{x} - \bar{\bar{X}})^2$	68.4	8.48	48.2	
$n(\bar{x} - \bar{\bar{X}})^2$	2873	365	1542	$\Sigma = \text{BSS} = 4780$
s^2	65.4	68.5	46.1	
$s^2(n-1)$	2682	2879	1429	$\Sigma = \text{WSS} = 6990$
$F_{(df_1 = k-1,\ df_2 = N-k)}$	$\dfrac{\text{BSS}/(k-1)}{\text{WSS}/(N-k)} = \dfrac{4780/2}{114/6990} = 39.0$			
p value	Almost 0			

in the numerator and $(N - k)$ degrees of freedom in the denominator, where k is the number of groups and N is the total number of observations ($= n_A + n_B + n_C + \cdots$)

$$F = \frac{\text{BSS} / (k-1)}{\text{WSS} / (N-k)}$$
$$= \frac{\text{variance of the group means}}{\text{average within-group variance}} \tag{6.26}$$

For the life expectancy data we obtain an F value of 39.0 with 2 and 114 degrees of freedom. The calculations are shown in Table 6.3, although in practice we would use our software package to obtain the answer.

Looking at the F statistic and its associated p value we conclude there are significant differences in life expectancy between the groups of countries. The null hypothesis of equal population means is rejected at greater than 99.9% confidence.

> **Key concept 6.7 Analysis of variance**
>
> An **analysis of variance (ANOVA)** is used to test whether three or more groups of data have the same population mean.
>
> The analysis looks at the **between-groups sum of squares (BSS)** and compares it with the **within-groups sum of squares (WSS)** to form an F statistic. This is used to assess whether the variation between groups is large relative to the variation within groups. If it is, it suggests that the differences between the groups are unlikely to have arisen by chance.

Measuring the effect

Another way of understanding ANOVA is to look at how it partitions the total variation in a data set into two parts – that which is due to error and that which is the effect of some differential treatment of the groups.

To talk of a treatment is to allude to medical trials where ANOVA is used to compare the effect of different treatments (also called interventions) on patients within each group, to assess which is the most effective. Here, however, it is intended in a more general sense: citizens of poorer countries are not 'treated' the same as those in more wealthy countries; we expect the different social, economic and cultural circumstances to have different 'interventions' upon the lives of the citizens and so have 'an effect' on life expectancy.

The total variation is measured as the total sum of squares and is itself the sum of the between-groups and within-groups sum of squares,

$$SS_{TOTAL} = BSS + WSS \tag{6.27}$$

If the treatment has an effect then it will create differences between the groups. However, differences will also arise for other (unexplained) reasons affecting the samples of data. It follows that Equation 6.27 can also be written as

$$SS_{TOTAL} = SS_{TREATMENT} + SS_{ERROR} \tag{6.28}$$

And the F statistic as

$$F = \frac{SS_{TREATMENT}/(k-1)}{SS_{ERROR}/(N-k)} \tag{6.29}$$

This makes clear that the F statistic is looking at how much of the variation in the data can be explained by the different treatments and is not just a consequence of the sampling.

It is useful to form standardised measures of the effect of the treatment, allowing different studies to be compared.

One measures is

$$\eta = \frac{SS_{TREATMENT}}{SS_{TOTAL}} = \frac{BSS}{BSS + WSS} \tag{6.30}$$

This gives an effect size of $\eta = 0.41$ for the country data. (The Greek symbol is eta.) Unsurprisingly, levels of national wealth have an effect on life expectancy.

Key concept 6.8 Measuring effects

The **effect** of a treatment (or intervention) is the difference it creates between groups; for example, when one group has received the treatment and another has not.

However, the difference may also arise as a chance consequence of how the groups were selected. The F test works by petitioning the total variation in the data into two parts – that which is explained by the treatment and that which is due to error.

By determining what proportion of the total variation is explained by the treatment, the strength of the effect can be measured. Small effects can be important but are harder to detect.

Hypothesis testing begins with the null hypothesis of no effect. If the true effect is small and so also is the sample size, the power of the test will be low (with conventional alpha values) making it unlikely that the null hypothesis will be rejected and leading to Type II errors.

As a rule of thumb, an effect size of between about 0.2 and 0.5 is small, of about 0.5 and 0.8 is medium and above 0.8 is large. These are relative terms, and it would be a mistake to believe a 'small effect' is therefore not of statistical or substantive interest. It could be. However, smaller effects are harder to detect, which takes us back to our earlier discussion about power (Section 6.4). If previous studies suggest that the effect is real but small, then it is wishful thinking to imagine a small sample will be able to detect it.

Contrasts

The ANOVA indicates the groups do not have the same population mean but does not show if any one group is especially different from the rest. By looking again at Figure 6.5 it seems the low-income group is different from the middle-income groups, whereas the lower and upper middle groups may be similar.

Contrasts are used to make specific comparisons of one or more of the groups with one or more of the others. For example, contrasting the low-income group with both the middle-income groups (together) gives a test statistic of $F = 70.8$ with 1 degree of freedom in the numerator, 115 in the denominator, and a p value that says we can be more than 99.9% confident that low-income countries have a mean life expectancy significantly lower than middle-income ones.

Contrasting the two middle-income groups with each other whilst omitting the low-income countries altogether gives a statistic of $F = 1.26$, also with 1 and 115 degrees of freedom but a p value of 0.26. There is no significant difference in life expectancy between the two groups of middle-income countries at conventional confidence levels.

In both these examples the use of contrasts creates the equivalent of a two-sample t test assuming equal variance, where $F = t^2$ and the p value is the same for either test.

Refer to the help pages of your statistical package for further information about how to set contrasts. In general, the contrasts sum to zero as in the following examples:

Group	A	B	C	D	
Contrasts	−1	1	0	0	Contrasts group A against group B
	−1	−1	1	1	Contrasts group A with B against C with D
	−2	0	1	1	Contrasts group A against C with D
	1	−3	1	1	Contrasts group B against A with C and D

6.8 Non-parametric tests

We noted in Section 6.6 that t tests are for samples drawn from normally distributed populations. Although some departure from perfect normality is not unreasonable, there are occasions when the assumption is not credible.

One such occasion is shown in Figure 6.6, which returns to the micro-data used in Chapter 3 and summarises the distances travelled by 20 randomly selected pupils (per

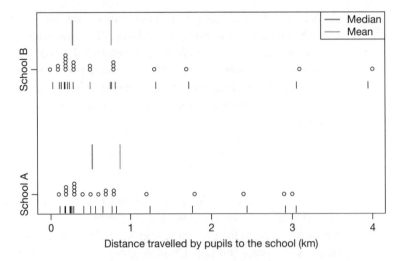

Figure 6.6 Strip charts and rug plots showing the distance travelled by pupils to one of two schools in London. The distribution is positively skewed, which the difference between the sample mean and median distances confirms

school) from home to one of two elementary schools in London. The main graphic is a **strip chart** (or dot plot) where each distance is rounded to the nearest 100 metres and plotted along the horizontal axis, stacking coincident values one above the other. The end result is similar to a histogram. Underneath the strip chart is a rug plot indicating the (non-rounded) values more precisely. The mean and median distances for each school are also shown.

The strip charts show the distributions to be positively skewed. The skew values are Sk = 1.38 for the school A data and Sk = 2.01 for school B. What the shape is showing is that pupils tend to travel to a near school, but uneven population distribution, the mismatch between the supply of and demand for school places, having a sibling already in the school, and a system allowing parents to express a preference for which school their child attends, create the long tail of pupils travelling from further afield (Harris and Johnston 2008; Sutherland *et al.* 2010). There is evidence of a distance decay effect where both the attractiveness of attending the school and the opportunity to do so diminish as the distance from home to school increases (which is more usual than the normal distribution for the one school in Chapter 3).

For data where the normality assumption cannot be met, one option may be to transform the data using one of the transformations listed in Chapter 3, Table 3.7. If this is unsuccessful or not possible, **non-parametric tests** offer an alternative.

Non-parametric tests are those that do not begin with fixed assumptions about how the population is distributed (or, more specifically, about the parameters that define the shape of the distribution). An alternative to the two-sample *t* test is the **Wilcoxon rank sum test**, also known as the Mann–Whitney test (Wilcoxon 1945; Mann and Whitney 1947).

The idea behind the test is to combine both samples, rank their values from 1 to (n_A + n_B) and then calculate the sum of the ranks given to members of the first sample. If the samples are drawn from the same population then we expect them to have similar distributions. Even permitting some difference due to sampling, both samples ought to contain a broadly equal mix of the high-, mid- and low-ranked values. If so, the first sample should not be dominated by especially high or low ranks and the sum of ranks for its members should be neither very high nor low.

The null hypothesis for the Wilcoxon–Mann–Whitney test is that the two samples have the same population distribution. The alternative hypothesis is that there is a 'location shift': that the median of one population is not equal to the other. The median is considered instead of the mean because there is no assumption that the populations are normally or even symmetrically distributed.

The p value for the test statistic can be calculated using what is known as a **permutation test**, calculating all possible outcomes and seeing what proportion of those exceed the test statistic. However, the number of outcomes rises rapidly with the sample sizes. For this reason an approximation may be used. The R software defaults to the approximation when the sample sizes exceed 50.

For the travel to school data, the Wilcoxon–Mann–Whitney test gives a p value of $p = 0.387$. This is insufficient to reject with 95% confidence the null hypothesis of equal median distance travelled by pupils to each of the two schools (the sample median is 534 m for school A and 295 m for school B).

An ANOVA assumes that the observations are independent of each other, that each group of data has equal variance and is normally distributed. When the assumptions of the ANOVA are violated, a non-parametric alternative is the **Kruskal–Wallis test**, which is a generalisation of the Wilcoxon–Mann–Whitney test.

Key concept 6.9 Parametric and non-parametric tests

Many statistical tests make an assumption about how the samples and/or their populations should be distributed. Often the assumption is of normality. These are **parametric tests**.

When the assumption is not warranted and the data cannot be transformed to make it so, a different sort of test is required. **Non-parametric tests** do not make assumptions about the shape of an underlying distribution and can be used to compare small or skewed data sets, or non-continuous data (for example, ranked values).

6.9 Conclusion: from detection to explanation

This chapter has presented methods to detect differences and effects, and to determine the probability that arose owing to change. When that probability is low (beneath the chosen alpha value), the effect is said to be statistically significant.

Whilst it is interesting to obtain statistically significant results, ultimately all the test is doing is describing a difference and not saying how it arose. For a medical trial, where one group of patients receives a new treatment and the other a placebo, we may reasonably assume the difference is caused by the treatment. We have an explanation for how it was created. Similarly, for the sort of agricultural trial described at the beginning of Chapter 5, with careful design of the experiment we can assume that whilst other factors (soil nutrition, microclimate, pests, etc.) will have an effect on the crop yield, there is also no reason to assume they are having a disproportionate effect on the plots of land where, for example, a new fertiliser is trialled. In this way, any increased yield on the plots with the fertiliser is likely to be caused by it.

However, social and environmental processes rarely unfold under controlled conditions whereby some people, places and species uniquely are exposed to an affecting influence whereas others are not. Nor do the effects of the processes act independently of each other. That makes explanation about 'what causes what' extremely difficult. However, if research (both quantitative and qualitative) is 'concerned with developing concepts, understanding phenomena and theoretical propositions that are relevant to other settings and other groups of individuals' (Draper 2004, p.645) then it is important we try. In the next chapter we look at some of the statistical tools available to do so.

Key points

- It is unsurprising if a sample mean is different from a hypothesised value or from the mean for another sample. Differences will arise by chance.

- What matters is the extent of the difference, relative to the variability of the data and giving consideration to the amount of data collected. We have more confidence in saying the difference is statistically significant if the difference is great, if there is little variability in the data and if there is a reasonable amount of data to found the inference upon.

- A formal process of hypothesis testing begins with the null hypothesis of no difference, of no effect. A statistical test is formed to measure the difference and an assessment of how unusual the result is – of how often it would be exceeded, in the long run.

- The act of rejecting the null hypothesis when it is correct is known as a Type I error and often the focus is on minimising this risk. However, that can lead to the reverse problem, of retaining the null hypothesis when it is wrong – a Type II error. The power of a test assesses the probability of rejecting the null hypothesis when it is right to do so: when there is, indeed, an effect; when it is not due to chance.

- The power of the test is low if the effect and/or the sample size is low. Essentially, the test has been designed to fail. Reciprocally, with the proliferation of large social and environmental data sets, we should not rush to label an effect as significant in any substantive or theoretical sense just because it appears to be statistical significant. If the set is large enough, even the most negligible differences will be confidently detected.

- To investigate whether a population mean could be equal to a hypothesised value, or whether two sampled populations have an equal mean, t tests are used. Analysis of variance (ANOVA) is used for three or more groups. If the distributional assumption of normality is not credible then it may be possible to transform the data or else to use non-parametric tests instead.

References

Cohen, J. (1988) *Statistical Power Analysis,* 2nd edn, Hillsdale, NJ: Lawrence Erlbaum.

Draper, A.K. (2004) The principles and application of qualitative research. *Proceedings of the Nutrition Society,* 63(04), 641–646.

Harris, R. and Johnston, R. (2008) Primary schools, markets and choice: studying polarization and the core catchment areas of schools. *Applied Spatial Analysis and Policy,* 1(1), 59–84.

Lenth, R. (2006) *Java Applets for Power and Sample Size* [Computer software], Available at: http://www.stat.uiowa.edu/~rlenth/Power, accessed 20 October 2010.

Mann, H. and Whitney, D. (1947) On a test of whether one of two random variables is stochastically larger than the other. *Annals of Mathematical Statistics,* 18(1), 50–60.

Murphy, K.R. and Myors, B. (2003) *Statistical Power Analysis: A Simple and General Model for Traditional and Modern Hypothesis Tests,* 2nd edn, Hillsdale, NJ: Lawrence Erlbaum.

Strahler, A.N. (1952) Hypsometric (area-altitude) analysis of erosional topography. *Geological Society of America Bulletin,* 63, 1117–1142.

Sutherland, R., Ching Yee, W., McNees, E. and Harris, R. (2010) *Supporting Learning in the Transition from Primary to Secondary Schools,* Bristol: University of Bristol.

Welch, B. (1949) Further note on Mrs. Aspin's tables and on certain approximations to the tabled function. *Biometrika,* 36, 293–296.

Wilcoxon, F. (1945) Individual comparisons by ranking methods. *Biometrics Bulletin,* 1(6), 80–83.

Relationships and explanations

Chapter overview

In previous chapters we looked at measures of variation around the centre of a data set and introduced hypothesis testing to infer if two or more sets of data are drawn from the same population. This chapter moves on to ask whether the variation we find in one variable is associated with the variation we find in another – to ask if they are related.

The chapter introduces scatter plots as a simple but effective way of detecting a relationship between two variables. The simplest is a straight-line relationship where the values in one variable either increase with the values in the other (a positive relationship) or increase as the other decreases (a negative relationship). Pearson's correlation coefficient, r, is a measure of the amount of straight-line association the two variables have.

A line of best fit can be added to the scatter plot to summarise the relationship. That line is introduced as a regression line. The logic of regression and the meaning of the various statistical tests and diagnostics presented by statistical software are explored and explained, for example F tests and goodness-of-fit measures. Multiple regression is considered, where more than one variable is used to explain another. The assumptions of regressions, how to test if they are met and what to do if they are not are reviewed.

Learning objectives

By the end of this chapter you will be able to:

- Draw a scatter plot to look at the relationship of two variables.
- Use Pearson's correlation coefficient to measure the linear association of two variables.
- Know the difference between a dependent and independent variable, and between a positive and negative relationship.
- Outline the principles of ordinary least squares (OLS) regression.
- Know how to interpret the regression outputs typically generated by statistical software.
- Know the assumptions of OLS, know how to check if they are met and understand the consequences if they are violated.
- Outline a strategy for undertaking regression analysis in a top-down way using goodness-of-fit measures, F tests and t tests to guide the decision-making process, identify insignificant variables and compare models.

7.1 Looking for relationships

Of the three uses of statistics mentioned in Chapter 1, we have discussed description and inference. Descriptive statistics allowed us to gauge the amount of variation around the centre of some data, and inferential statistics helped us to judge where the true centre might be. In this chapter we turn to relational statistics to see if the variation we find in one set of data corresponds to the variation we find in another.

From a geographical perspective we might ask if the patterns seen in one map correspond with the patterns we see in another. The left panel of Figure 7.1 is a choropleth map of average property price for 35 neighbourhoods within a city, and the right panel shows the percentage of the population with qualifications above high school. In a market-based economy where education attracts a wage premium and wages afford property, we expect the maps to show similar patterns and broadly they do. Of course, our understanding of the maps is affected by how the data are grouped and by how the maps are shaded (Monmonier 1996). In this example the two variables are each split into quartiles with lighter shading indicating lesser values and each group containing about one-quarter of the neighbourhoods.

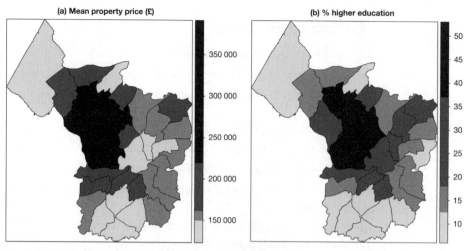

Figure 7.1 Maps indicating the mean property price and the percentage of the population with higher education qualifications for neighbourhoods within Bristol, UK. They have similar patterns, suggesting the variables are related

7.2 Using scatter plots

Table 7.1 gives the data behind the maps. To aid interpretation, the data are sorted into ascending order by the mean property price per neighbourhood. Even so, it is not easy to discern the relationship between property prices and education. It is much easier if the data are plotted on a scatter plot.

A **scatter plot** is a useful tool for seeing if two variables are related and for detecting observations that are distorting the relationship in some way. To draw one, we place a symbol on the plot at the coordinate representing each pair of data values. For

Table 7.1 The property price and higher qualifications data for neighbourhoods within a city

Obs	X	Y	Obs	X	Y	Obs	X	Y	Obs	X	Y
9	6	123 080	28	14	151 469	5	17	172 058	22	40	281 351
23	11	135 687	1	7	151 741	11	18	176 502	18	51	301 033
33	24	139 678	2	14	156 004	31	25	177 319	14	51	323 527
4	6	140 676	13	14	157 999	21	14	180 856	19	49	349 739
35	9	141 220	3	13	158 906	6	33	183 395	16	35	391 189
7	7	141 492	12	9	159 723	27	43	217 952			
34	16	143 760	29	25	160 267	15	39	220 220	$\bar{x}$	24	192 592
8	8	148 476	24	14	161 174	20	45	238 450	s	16	670 72
25	20	148 839	10	30	164 076	17	53	259 402			
26	12	150 562	30	18	171 695	32	52	261 216			

Obs = observation number (position in data set prior to sorting).

Y = neighbourhood mean property price (£).

X = percentage of population aged 16–74 with higher level qualifications (above A-level).

Source: *Venue magazine*, No. 857 (2009), from various other sources.

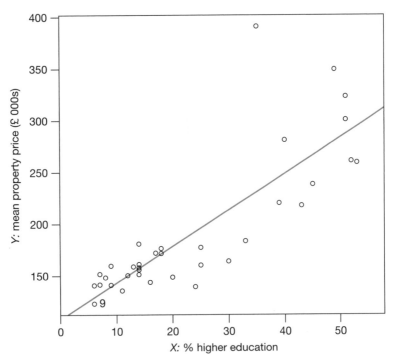

Figure 7.2 Scatter plot showing a positive relationship between the average property price and the rate of higher qualification within neighbourhoods. A line of best fit also is shown

example, observation 9 in Table 7.1 is represented by the circle at position ($x = 6$, $y = $ 123 080) in Figure 7.2.

A **line of best fit** has been added to the scatter plot. How to fit that line is described later in the chapter. For now we can describe the relationship between property prices and higher education qualifications as positive in a purely mathematical sense: the line is upward sloping; it has a positive gradient. If the line had been downward sloping, it would be a negative relationship.

> **Key concept 7.1   Scatter plots**
>
> **Scatter plots** are a simple but effective way of seeing the relationship between two variables and to detect any errors prior to further analysis. The horizontal axis represents the X variable and the vertical axis the Y variable.

7.3   Independent and dependent variables

It is conventional for the variable that leads to or 'explains' the other to be represented by the horizontal axis on a scatter plot, known as the X variable. The second variable is plotted on the vertical axis and is the Y variable. **The X variable is often referred to** as the independent variable and the *Y* variable as the dependent variable. The terms predictor and response variable are also used.

In some cases the order of the variables may be obvious. The number of rainy days in a month has an effect on umbrella sales, but umbrella sales cannot create rain. The amount of rainy days will be the *X* variable and umbrella sales will be the *Y* variable.

In other cases the direction of the relationship may be unclear or debatable. Figure 7.2 is drawn with the assumption that those who have higher educational qualifications tend to attract a higher salary and therefore bid up property prices in the more desirable neighbourhoods. Hence the percentage of higher qualifications is the *X*; average property price is the *Y*. However, the relationship could actually work the other way around with neighbourhoods of higher property price being affordable only to those with higher qualifications.

Some relationships are direct. The price of an airline ticket increases with demand for the flight. Others are indirect. Wilkinson and Pickett (2009) suggest a relationship between (greater) income inequality and (lower) levels of recycling for more wealthy countries. Does income inequality create more non-recycled waste? No, but the two may be linked by an underlying sense of public responsibility and social cohesion.

It also is important to recognise the distinction between analysis undertaken at one scale and inferences made at another. For our case study the scatter plot implies that more highly qualified persons purchase more expensive property. However, that inference is not necessarily supported by the data. What the plot actually shows is that *neighbourhoods* with higher percentages of higher qualified persons are also the neighbourhoods that have higher average property prices. It does not follow that the higher qualified *persons* have to be living in those expensive properties. They could be living in cheap housing within otherwise affluent neighbourhoods.

To assume a relationship measured at one geographical scale (here, neighbourhoods) necessarily is true at another scale (individuals) is to risk what is known as the **ecological fallacy** (see Chapter 8). What might be regarded as an unfortunate mistake risks becoming pernicious if, for example, you observe that neighbourhoods with high crime rates are also those that suffer from poverty and you use that finding to 'prove' there is a 'criminal' or 'deviant' mentality amongst the least economically wealthy (an argument that some writers about a social 'underclass' do make).

Key concept 7.2   Independent and dependent variables

It is conventional to assume that what is measured as the *X* variable leads to or explains what is measured as the *Y* variable. As such, the *X* variable is known as the **independent** (or predictor) variable, whereas the *Y* variable is the **dependent** (or response) variable. Whilst this implies *X* causes *Y*, in practice their relationship may be indirect or uncertain.

7.4 Correlation

Two variables that appear related may be described as correlated. A **correlation coefficient** is a statistic that describes the degree of association between two sets of paired values (Hammond and McCullagh 1978).

We have seen that as the percentage of persons with higher education qualifications increases so generally does the mean property price in neighbourhoods within the study region. However, it is not true to say that these X and Y variables always rise together. If we redraw Figure 7.2 to show, in Figure 7.3, the mean of both the X and Y variables, we see there are neighbourhoods that are above the mean for higher qualifications but below the mean for property prices. (These are the sorts of places where university instructors live!)

To assess how correlated the variables are we begin by calculating their **covariance**. To do so we take each observation in turn, look at its pair of x and y values, calculate how far each is from the mean for that variable, and multiply those deviations together. Next we sum the results for all pairs and create an average by dividing by the degrees of freedom.

In mathematics the answer obtained by multiplying two values together is called the product. Hence, the covariance is the sum of products divided by the degrees of freedom:

$$\text{cov}(X,Y) = \frac{\sum_{i=1}^{n}\left(x_i - \bar{x}\right)\left(y_i - \bar{y}\right)}{n-1} \tag{7.1}$$

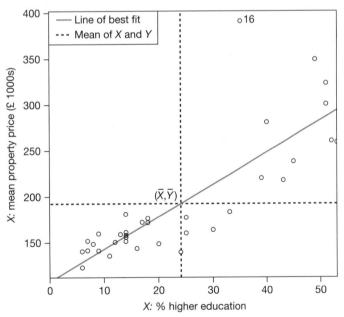

Figure 7.3 Whilst the general relationship is positive, the relationship is not perfect. There are observations that are *above* the mean for rates of higher education but *below* the mean for property price. Observation 16 is an outlier (see Section 7.5, below)

The problem with the covariance statistic is that the result reflects the measurement units of the data. If we measure property price in pounds sterling and rates of higher educational attainment in percentages, the covariance for the neighbourhood data is approximately 859 771. If, however, we convert the property prices into US dollars and the percentages into proportions, the covariance is about 14 100. This dependence on the measurement units is unhelpful if we are trying to make comparisons across data sets. The solution is to standardise the data by converting each pair of observations into z values (see Chapter 3, Key concept 3.6):

$$z_x = (x - \bar{x})/s_x \qquad z_y = (y - \bar{y})/s_y \qquad (7.2)$$

Equation 7.1 then gives what is called the correlation coefficient, r,

$$r = \frac{\sum_1^n (z_x - \bar{z}_x)(z_y - \bar{z}_y)}{n - 1} \qquad (7.3)$$

which, because the mean of each set of z values is zero (by definition), simplifies to

$$r = \frac{\sum_1^n z_x z_y}{n - 1} \qquad (7.4)$$

More specifically, r measures **Pearson's product-moment correlation coefficient**, named after the statistician Karl Pearson (1857–1936) with initial development by Sir Francis Galton (1822–1911). Pearson's correlation coefficient for the higher qualification and property price variables is $r = 0.821$ and remains so whether the rate of higher qualification is by percentage or by proportion, and whether the property price is in pounds or dollars.

The possible values for the coefficient range from $r = -1$, a perfect, negative correlation, to $r = 1$, a perfect, positive correlation. 'Perfect' means one variable could be used as a direct substitute for the other with all observations lying exactly on the line of best fit, as in the top row of Figure 7.4. A value of $r = 0$ indicates zero correlation, implies no relationship and causes the line of best fit to be flat (final graph in Figure 7.4).

In practice, measurement error and the complexity of the real world mean that finding a perfect relationship is unlikely. It may instead indicate that X and Y are really the same variable expressed in different measurement units. Each of the following gives a perfect correlation but only because the X and Y are equivalent in each case:

- Property prices measured in pounds sterling (X) and property prices measured in dollars (Y) (gives $r = 1$).
- Percentage of the population with higher qualifications (X) and proportion of the population with higher qualifications (Y) (gives $r = 1$).

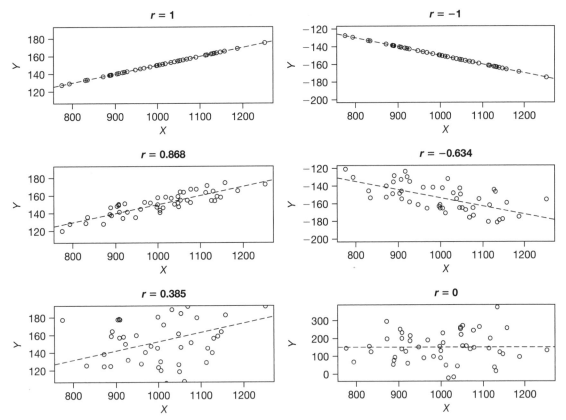

Figure 7.4 Positive and negative correlations of various strengths. For a perfect relationship where $r = \pm 1$ all of the points lie on the line of best fit. When there is no relationship, $r = 0$ and the line of best fit is flat

- Percentage of the population with higher qualifications (X) and the proportion of the population without higher qualifications (Y) (gives $r = -1$).

Testing the statistical significance of the correlation coefficient

The statistical significance of the correlation coefficient can be assessed with a hypothesis test (see Chapter 6 for an introduction to hypothesis testing). The null hypothesis is that the true correlation is equal to zero, and the alternative hypothesis is the opposite. The symbol ρ (rho) is used to differentiate the true correlation from the sample-dependent correlation coefficient, r:

$$H_0 : \rho = 0$$
$$H_1 : \rho \neq 0 \tag{7.5}$$
$$\alpha = 0.01$$

Provided each variable is normally distributed or the sample size not very small, the test is a t test using the equation

$$t = \frac{r}{\sqrt{\dfrac{1 - r^2}{n - 2}}} \tag{7.6}$$

For our data set, where $r = 0.821$ and $n = 35$, this gives a value of $t = 8.26$ with $n - 2$ degrees of freedom. Knowing this, from a statistical table we obtain a p value of $p < 0.001$. From it, we can be (more than) 99% confident that the true correlation is not zero.

Looking at Equation 7.6 it can be seen that whether the null hypothesis is rejected (or not) is a function not only of the correlation between X and Y, but also of the sample size. Together these determine the power of the test (Chapter 6, Section 6.4). When $n = 35$ any correlation of $r > 0.430$ or $r > -0.430$ will be significant at a 99% confidence level. When $n = 100$ the threshold is $r = \pm 0.256$ and for $n = 1000$ it is only $r = \pm 0.081$.

Gauging the strength of an effect therefore is different from testing if it is statistically significant or not. A small effect can be significant given enough data and a large effect insignificant if the data are few. As a rule of thumb, a correlation of about $r = \pm 0.1$ can be considered as a small effect, $r = \pm 0.3$ as medium and $r = \pm 0.5$ as large (Rice and Harris 2005). However, the strength of any effect should be interpreted in the context of what is being studied and what previous research has shown.

7.5 Complications with the correlation coefficient

Non-linear relationships

Pearson's correlation coefficient, r, measures the linear correlation of two variables. It assumes their relationship can be summarised in terms of a straight line of best fit. Not all relationships necessarily are that simple.

For example, the relationship between property price and the amount of green space appears U-shaped in Figure 7.5. What then happens is the more negative relationship on one side of the curve is balanced by the more positive relationship on the other. The end result is an r value close to zero.

It is sometimes possible to straighten a curved relationship by applying one of the transformations listed in Chapter 3, Table 3.7, to either the X or Y variable, or both. This will be helpful when the data are skewed.

Group effects

Figure 7.6 gives a second example of when Pearson's correlation coefficient is deceptive. The correlation is $r = 0.940$, indicating a positive relationship. However, there

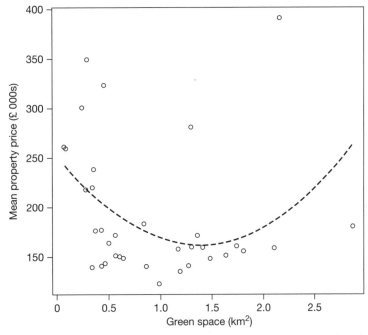

Figure 7.5 The property price and green space variables are not linearly related

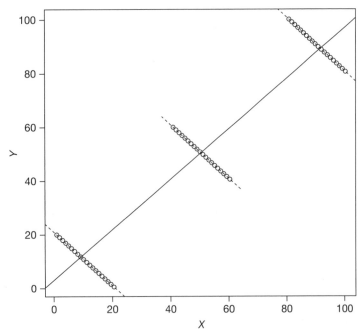

Figure 7.6 An example of when Pearson's correlation coefficient and the line of best fit are deceptive. They both indicate a positive relationship ($r = 0.940$) but the relationship actually appears to be negative for three groups of data

clearly are three distinct groups of data and they each show a negative relationship. The scatter plot has helped to identify the problem.

Outliers

Pearson's correlation coefficient is calculated using the mean and standard deviation of each variable, neither of which is robust in the presence of outliers (see, for example, Chapter 2, Figure 2.8).

An outlier can be defined as an observation that is a long way above or below the line of best fit, relative to all the other data points. In that sense it is unusual because it does not fit the general trend and should, perhaps, be removed from the data set and/ or treated as an exceptional case.

We have identified observation 16 as an outlier in Figure 7.3. Its effect is to exaggerate the amount of variation around the mean of the Y variable but also to decrease how that variation relates to the variation in the X variable. The initial correlation is $r = 0.821$ but rises to $r = 0.891$ if the observation is omitted, an increase of almost 10%.

Non-parametric tests

Use of Pearson's correlation coefficient assumes both variables are approximately normal. In circumstances when this assumption clearly is violated but the data can be ranked from lowest to highest for each variable, **Spearman's rank correlation coefficient** or Kendall's rank correlation coefficient may be used instead. For further details see, amongst others, Hammond and McCullagh (1978, Ch. 7) or Robinson (1998, Ch. 4).

Key concept 7.3 Pearson's correlation coefficient, *r*

If the values in one variable appear to rise or fall with the values of another then the two are correlated. A **positive correlation** is where the two sets of values rise together, whereas a negative correlation is where one variable increases as the other decreases.

Pearson's correlation coefficient measures the linear relationship of two variables. It ranges between −1 for a perfect negative relationship and +1 for a perfect positive relationship. A value of zero implies the variables are unrelated.

7.6 Bivariate (two-variable) regression

In the preceding sections we talked about a line of best fit drawn on the scatter plots. Specifically it is a **regression** line, taking its name from the studies Sir Francis Galton undertook exploring the relationship between the heights of parents and the heights of their children, and his discovery of 'regression to the mean' (or 'regression toward mediocrity' as he actually described it) (Hepple 2001).

A regression line is used to model the relationship between two variables, describing it in terms of a straight line. A line is used because it is simple and therefore a sensible way to begin.

The equation of a straight line

Figure 7.7(a) illustrates a positive relationship where an increase in Y of 25 units is related to an increase of 50 units in X. If we denote the change in Y as Δy and the change in X as Δx, then

$$\frac{\Delta y}{\Delta x} = \frac{25}{50} = 0.5 \tag{7.7}$$

Equation 7.7 calculates **the gradient** of the line (its slope) and suggests a 1 unit change in x has the effect of raising y by 0.5 units.

In Figure 7.7(b), the line is downward sloping and has a gradient of

$$\frac{\Delta y}{\Delta x} = \frac{-50}{25} = -2 \tag{7.8}$$

It is a negative relationship. For every 1 unit increase in x we expect a 2 unit decrease in y.

The gradient relates change in the X variable to change in the Y variable but is insufficient to define a straight line. We need also to know the **y intercept**: the value of y when $x = 0$. In some cases, we would expect it to be zero. If there are no people, there can be no deaths. In other cases we expect some amount of the Y variable before the effect of the X variable. For example, granite is a natural source of radiation,

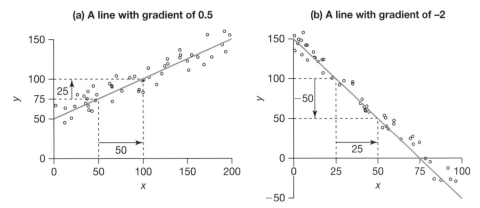

Figure 7.7   Two regression lines. In (a) a 1 unit increase in x has the effect of increasing y by 0.5 units. In (b) the 1 unit increase decreases y by 2 units

containing uranium that decays to form radon gas and is carcinogenic. We therefore expect a relationship between the amount of exposure to radon and the probability of being diagnosed with lung cancer. However, radon is not the only cause of cancer and the probability of being diagnosed is not zero even when the amount of exposure is.

It is always sensible to check the value of the y intercept and whether it is plausible. If not, it may suggest the regression line has been fitted incorrectly.

Once the gradient and the y intercept are known, the line is defined by an equation of the form

$$y = \text{intercept} + (\text{gradient} \times x) \tag{7.9}$$

which can also be written as

$$y = b_0 + b_1 x \tag{7.10}$$

where b_0 represents the intercept value and b_1 is the gradient. It is sometimes written as $y = a + bx$ (for example, in Freedman 2009; Rogerson 2006), or as $y = mx + c$ (in this case putting the gradient before the intercept, as in Bostock and Chandler 2000). The notation varies by author.

Looking at Figure 7.7(a), when $x = 0$, $y = 50$, which gives the y intercept. Now knowing that $b_0 = 50$ and that $b_1 = 0.5$, the equation of the line is

$$y = 50 + 0.5x \tag{7.11}$$

Be careful when reading the y intercept from a scatter plot. First check that the y axis is positioned vertically at $x = 0$. It will not necessarily be so: on most occasions in this chapter it is not (the graphs look better that way – it removes empty space). However, in Figure 7.7(b) it is. Use the plot to confirm the equation of the line is

$$y = 150 - 2x \tag{7.12}$$

Key concept 7.4 The equation of a straight line

The equation of a straight line can be written in the form $y = b_0 + b_1 x$ where b_0 is the y intercept (the value of y when $x = 0$) and b_1 is the gradient (how much y changes given a 1 unit change in x).

Finding the position of the regression line

We can now define a regression line by its y intercept (b_0) and its slope (b_1). **We may interpret the slope as measuring the effect** of X on Y and recognise the y intercept as

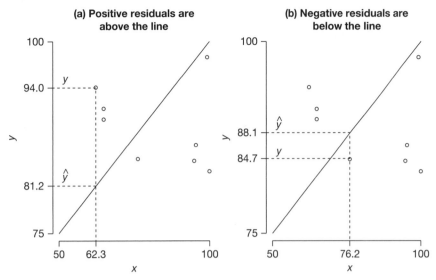

Figure 7.8 A residual is the difference between a y value observed in the data and the value suggested by the regression line, that is $(y - \hat{y})$

the amount of the Y variable when there is none of X. Even so we still need to know which is the line of best fit.

To begin, look at Figure 7.8, an enlargement of part of Figure 7.7(a), and note that the regression line is one of best fit but not perfect fit, because not all the data points are located on it; in fact, none is. Therefore, the measured values of Y differ from what the regression line predicts.

The differences between the observed (y) and expected values ($\hat{y}$) are called the residual differences: $(y - \hat{y})$. They are residual because they are left unexplained having fitted the regression line (the line says nothing about why the differences exist). They are also called the **residual errors**.

Residuals can be positive or negative. A positive residual is when the observed value exceeds the predicted value. A negative residual is when the observed value is less than the predicted value.

To measure the total deviation of the residuals above and below the regression line, a sum of squares calculation is used. It measures the sum of the squared error, SSE:

$$SSE = \sum_{i=1}^{n} \left(y_i - \hat{y}_i \right)^2 \tag{7.13}$$

The line of best fit is the one of all lines that could be drawn on the scatter plot, that minimises SSE. It gives

$$MIN \sum_{i=1}^{n} \left(y_i - \hat{y}_i \right)^2 \tag{7.14}$$

To locate the line, we could use a computer to work through the options and determine a solution. For more complex data structures and models that essentially is the approach used. Here, however, the line is the one where

$$b_1 = \frac{\sum_{i=1}^{n}(x_i - \bar{x})(y_i - \bar{y})}{\sum_{i=1}^{n}(x_i - \bar{x})^2} \qquad (7.15)$$

(The proof is in Crawley 2005, p.129, for example.)

Once the gradient, b_1, is known, the y intercept, b_0, can be calculated from knowing

$$\bar{y} = b_0 + b_1\bar{x} \Rightarrow b_0 = \bar{y} - b_1\bar{x} \qquad (7.16)$$

(cf. Equation 7.10). This is because the line goes through the mean of both the X and Y variables (the point $(\bar{x}, \bar{y})$ in Figure 7.3).

This type of regression is called ordinary least squares (OLS). In practice, b_0 and b_1 are determined not with Equations 7.15 and 7.16 but using matrix maths (see, for example, Freedman 2009; O'Sullivan and Unwin 2002).

Key concept 7.5   A residual

The regression line predicts a y value given a value of x. The predicted value is $\hat{y}_i = b_0 + b_1 x_i$.

A **residual** is the difference between the measured y value (found in the data) and what the regression line predicts the y value should be. It is $y_i - \hat{y}_i$.

A **positive residual** means the measured value is greater than the predicted value. A **negative residual** means the measured value is less.

7.7 Interpreting regression analysis

When a regression line is fitted to two variables we are suggesting their relationship can be understood in terms of a model of the form

$$y = \beta_0 + \beta_1 x + \varepsilon \qquad (7.17)$$

We are saying it is a straight-line relationship and are using our data to estimate the 'true effect' of X on Y, which is denoted by β_1. The y intercept is β_0 and we are admitting the unlikelihood of a perfect relationship by including an error term, ε.

Our best estimate of β_1 is the gradient of the regression line, b_1. When there is no effect the line should be flat and then b_1 is zero. This is the case in the last diagram of Figure 7.4 where the variation in the Y variable is random and entirely unrelated to X.

Measuring the gradient of the regression line therefore is the first stage of determining whether X has an effect on Y, of whether the variables are related. However, this alone is not sufficient because a non-zero gradient could arise by chance. Your statistical software will therefore test a null hypothesis of no effect, of the true gradient being equal to zero ($H_0 : \beta_1 = 0$). It will then give the p value, the probability of rejecting the null hypothesis when it is correct (a Type I error).

It does so by calculating a t statistic:

$$t = \frac{b_1}{se_{b_1}}$$ (7.18)

where the standard error is, as in previous chapters, a measure of uncertainty, now based on the average residual error (the numerator in Equation 7.19, below) and the variation of the X variable around its mean (the denominator):

$$se_b = \sqrt{\frac{\sum_{i=1}^{n}\left(y_i - \hat{y}_i\right)/(n-2)}{\sum_{i=1}^{n}\left(x_i - \bar{x}\right)^2}}$$ (7.19)

If the probability of exceeding the magnitude of the t value by chance alone is less than $p \leq 0.05$ we can be 95% confident that the effect of X on Y is not actually zero. If $p \leq 0.01$ we can be 99% confident.

Key concept 7.6   Bivariate regression

If two variables have a straight-line relationship (or can be transformed to have one), that relationship can be summarised by a model of the form

$$y = \beta_0 + \beta_1 x + \varepsilon$$

where β_1 is how much the y value is expected to rise for a 1 unit increase in x, and β_0 is the y value expected when $x = 0$. The error term, ε, recognises that the relationship is unlikely to be perfect.

The regression coefficients (β_0 and β_1) are estimated from a line of best fit, fitted to the data. The slope of that line (b_1) provides the estimate of β_1 which is the effect of X on Y. The y intercept, b_0, estimates β_0. The line is the one that minimises the sum of the squared deviations above or below that line, the sum of the squared residuals, that is $\text{MIN} \sum_{i=1}^{n}\left(y_i - \hat{y}_i\right)^2$.

β_1 is also known as **the effect** of X on Y. If $b_1 = 0$ the line is flat and the inference is $\beta_1 = 0$ too: the variables are not related. If $b_1 > 0$ it implies $\beta_1 > 0$ and that X has a positive effect on Y (it is a positive relationship). If $b_1 < 0$ it suggests $\beta_1 < 0$ and that the relationship is negative. However, the observed effect could be due to chance, a peculiarity of the data sample. Therefore, a hypothesis test is undertaken with the null hypothesis of the true effect being zero. If the null hypothesis is rejected at a suitable level of confidence it suggests the relationship between X and Y is statistically significant.

An example

Table 7.2 shows a regression analysis of the property price and higher qualifications data given in Table 7.1 (and in Figures 7.2 and 7.3). Under the column headed 'Estimate' we find the y intercept and the gradient of the regression line. The equation of that line is

$$\hat{y} = 107\,843 + 3523x \qquad (7.20)$$

This **predicts** that for every 1 unit increase in the proportion of the population with higher educational qualifications the average property price for the neighbourhood will increase by £3523.

Dividing the estimate of the slope (β_1) by the standard error gives a t value of 8.25. The probability that this result arises by chance is less than $p = 0.001$, so we can be greater than 99.9% confident that there is a statistically significant relationship.

In a similar way, the statistical package also has determined the probability that the y intercept (β_0) could be equal to zero. It is unlikely ($t = 8.838$; $p < 0.001$), which is logical. If it were zero it would imply properties are free to buy in areas without residents of higher qualification.

Other parts of the table are explained below.

The R^2 measure

The third row of Table 7.2 indicates how successful the regression line is at explaining the variation found in the Y variable. It gives a measure of **goodness of fit** – of how well the regression line fits the data.

The R^2 value is known as the coefficient of determination but more usually called **R-squared**. As the notation suggests, the R^2 value (for bivariate regression) is equal to the square of Pearson's r. The interpretation is therefore the same except the R^2 makes no distinction between a positive or negative relationship and ranges from zero (no relationship) to one (a perfectly positive *or* negative relationship). Like the correlation coefficient, using the R^2 value assumes we are measuring a straight-line relationship. A low R^2 may not be because of *no* correlation, just no *linear* correlation.

The F test (ANOVA)

The fourth row of Table 7.2 gives the result of an F test and the probability that it has arisen by chance.

Table 7.2 Summary statistics for a regression analysis of the relationship between mean property prices and the percentage of the population with higher educational qualifications in neighbourhoods in Bristol

	Estimate	Standard error	t	p
β_0 (y intercept)	107 843	12 202	8.84	<0.001
β_1 (slope)	3523	427	8.25	<0.001
R^2	0.677			
$F_{(1,33)}$	68			<0.001

The last time we encountered an F test was in Chapter 6, Section 6.7, where it formed part of an analysis of variance (ANOVA). There, the total variation in the data was partitioned into two parts: that found between groups, measured by the between-groups sum of squares (BSS); and that found within groups, measured by the within-groups sum of squares (WSS). The F test was used to determine whether the variation between groups was significant relative to the variation within groups.

The variation in a Y variable can also be partitioned into two: that which is explained by the regression line and that which is not. The F test then determines whether the amount explained is significant relative to the amount left unexplained.

To better understand how, we begin with some definitions. The total variation in the Y variable will be a sum of squares measurement: the sum of the squared deviations around the mean,

$$SSY = \sum_{i=1}^{n} \left(y_i - \bar{y} \right)^2 \tag{7.21}$$

Some of that variation is explained by the regression line, which predicts how far above or below the mean of the Y variable an observation will be, given its value for x. The predicted value is $\hat{y}_i$, so for any observation the regression line explains $(\hat{y}_i - \bar{y}_i)$. Summing for all observations, the total variation explained by the regression line is also expressed as a sum of squares measurement,

$$SSR = \sum_{i=1}^{n} \left(\hat{y}_i - \bar{y} \right)^2 \tag{7.22}$$

What is left unexplained is any difference between an observation's actual value and the value the regression line predicts. This is the residual error, $(y_i - \hat{y}_i)$. The total unexplained variation is

$$SSE = \sum_{i=1}^{n} \left(y_i - \hat{y}_i \right)^2 \tag{7.23}$$

SSE is the residual standard error shown in Table 7.2. Together, SSR and SSE sum to the total variation,

$$SSY = SSR + SSE \tag{7.24}$$

and the R^2 statistic is the proportion of the total variation explained by the regression:

$$R^2 = \frac{SSR}{SSY} \tag{7.25}$$

For bivariate regression, the F statistic is calculated as

$$F_{(1, n-2)} = \frac{SSR/1}{SSE/(n-2)} \tag{7.26}$$

From this, and knowing it has one and $(n - 2)$ degrees of freedom, a p value is obtained giving the probability that the statistic arose by chance. If that probability is sufficiently low it suggests the regression model goes a significant way in explaining the variation of the Y variable around its mean.

7.8 Assumptions of regression analysis

Ordinary Least Squares (OLS) estimates of β_0 and β_1 are sometimes described as **BLUE: Best Linear Unbiased Estimates**.

They are unbiased because, if you sampled often enough, fitted enough regression lines, and looked at the distribution of your b_0 and b_1 values, they would be centred on the true values β_0 and β_1, respectively. This is analogous to how the distribution of means from random samples will be centred upon the true population mean, μ, in the long run (see Chapter 5, Section 5.5).

They are best in that no other (linear and unbiased) estimates would have a smaller standard error. They therefore maximise our certainty in the data (remembering that the standard error is a measure of uncertainty).

However, OLS estimates will not be BLUE if certain conditions of the model are not met. An obvious requirement is that the model is linear. Perhaps confusingly, this does not preclude modelling a curvilinear relationship if one or both of the variables can be transformed to give a straight line. The linearity is in the parameters, β_0 and β_1. This means a model of the form $y = \beta_0 + \beta_1 x^2 + \varepsilon$ could be estimated using OLS but a model of the form $y = \beta_0 + \beta_1^2 x + \varepsilon$ could not.

Other conditions pertain to the residuals. These should be **independent and identically distributed (i.i.d.)**, which means they should be normally distributed as random deviations around the regression line. Their values should be independent of each other and have no unexplained pattern or structure. There should be no significant outliers.

Finally, the model assumes X leads Y, not the other way around, with no feedback from Y back to X.

7.9 Checking the residuals

To check that the residuals meet the assumptions of regression analysis outlined above, statistical software offers a range of visual tools, possible examples of which are shown in Figure 7.9.

The first plot shows the residual values for each of the observations: the difference between their actual values and what the model predicts, $(y_i - \hat{y}_i)$. It is clear that there is one positive residual that is much greater than the others, potentially affecting the line of best fit. It is for observation 16.

The second plot is a histogram, used to look at the distribution of the residuals and check they are normal. However, a better plot to do the same is the quantile plot, Figure 7.9(c) (Chapter 3, Section 3.11). Broadly the residuals are normally distributed, though two observations are creating a positive skew (16 again, and 19).

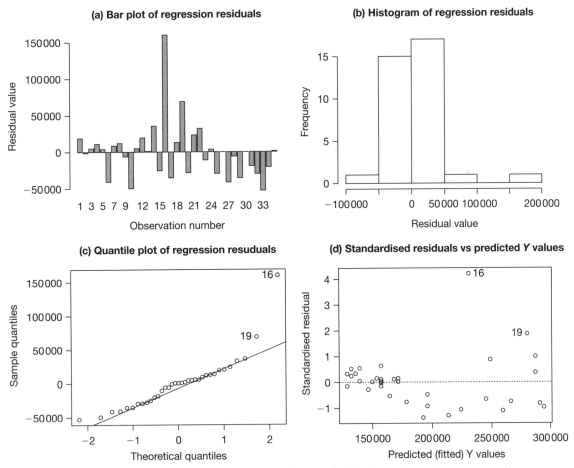

Figure 7.9 Examples of plots that can be used to check that the distribution of the residuals meets the assumptions of the regression analysis (normally distributed with no extreme residuals or non-random patterning)

Like (a), Figure 7.9(d) plots the residual differences but this time as a scatter plot against each observation's predicted y- value ($\hat{y}_i$). If the regression line perfectly fitted the data each observation would lie on the regression line and therefore the residual value would be zero in each case. In such a circumstance each of the points would lie on the horizontal zero line shown in Figure 7.9(d).

That they do not is not surprising. Some residual error is expected. However, the assumption of regression analysis is that it is **random error**, meaning there should be no patterning or unexplained structure. Looking at Figure 7.9(d), the greater the modelled y value, the greater the residuals appear to be (even ignoring observation 16 as an outlier). This suggests the residuals are not random but are dependent in some way upon the independent variable, X (because it is from X that we get the predicted value of Y: $\hat{y}_i = b_0 + b_1 x_i$).

Note that it is the **standardised residuals** that are shown on the vertical axis of Figure 7.9(d). We could have used the original ('raw') residuals and the plot would look the same. However, the standardised residuals have the advantage of being measured on a standardised scale that allows comparison across data sets independent of the measurement units. For bivariate regression the standardised residuals are calculated as

$$\text{standardised residual} = \frac{(y_i - \hat{y}_i)}{\sqrt{\text{SSE} /(n-2)}} \tag{7.27}$$

(cf. Equation 7.23).

As a rule of thumb, a standardised residual greater than two or less than negative two is a concern, and one greater/less than four/negative four even more so. Hence, though we may be cautious about observation 19, we should have particular concerns about observation 16: it is an **extreme outlier** and has three effects on the model. First, it affects the positioning of the regression line. Secondly, it is increasing the total variation of the observations around the regression line, increasing the uncertainty in the data (raising the standard error). Thirdly, it decreases the R^2 value.

For geographical data, the assumption that the residuals are random and without patterning applies to how they should appear when mapped. The assumption is suspect for our model's residuals where negative residuals tend to be concentrated in the centre of the city and positive residuals towards its edges (Figure 7.10).

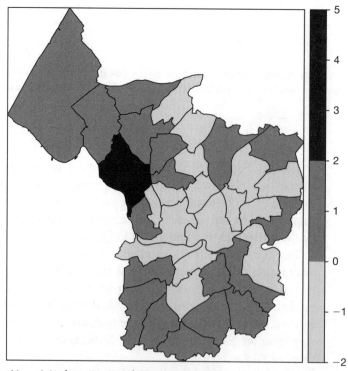

Figure 7.10 Map of the (standardised) residuals from the regression model. Negative residuals appear more in the centre of the city and positive residuals towards the outskirts

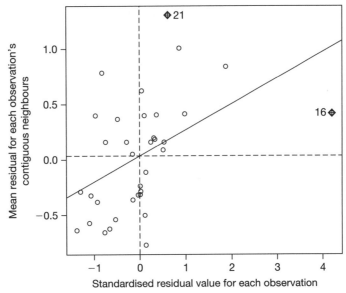

Figure 7.11 Moran plot comparing the (standardised) residual of each observation with those of its surrounding neighbours. The upward-sloping line indicates positive spatial autocorrelation

In addition to the map, a useful way of revealing any geographical trends is to compare the residual value for each neighbourhood with the average for its contiguous neighbours (the neighbourhoods with which it shares a border). This is what a **Moran plot** does and Figure 7.11 is an example. The general trend is for a positive residual in one part of the map to be surrounded by other positive residuals, and a negative residual in another part of the map to be surrounded by other negative residuals. In terms of the residuals, then, near values are alike – a pattern of positive spatial autocorrelation in the data (see Chapter 1, Key concept 1.5). This suggests the residuals are neither random nor independent of each other as they should be.

Standard statistical software does not produce maps or allow Moran plots to be produced. However, some GIS software has statistical capability, and other software is available that integrates statistical and spatial thinking. These include GeoDa (http://geodacenter.asu.edu/) and the sp and spdep libraries for R (Bivand *et al.* 2008).

Leverage points

An observation that has values a long way above or below the mean of both the *X* and *Y* variables can act like a lever on the regression line. Figure 7.12 gives an example of the problem.

Cook's distance commonly is used as a measure of each observation's influence upon the line. Any observation for which Cook's distance is close to or exceeding one, or is much greater than for other observations, should be looked at (Maindonald and Braun 2006).

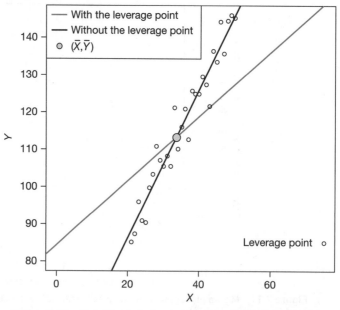

Figure 7.12 The effect of a leverage point on a regression line

In our model of property prices the greatest Cook's distance is for observation 16 with a score of 0.4. This is double the next highest score, which is for observation 19 (cf. Figure 7.9).

Key concept 7.7   Outliers and leverage points

An **outlier** is an observation that is seen on a scatter plot as a long way above or below the regression line in comparison with other data points. It is, in that sense, different from the rest and consideration should be given to whether it should be included in the model and/or treated as categorically different from the rest.

A **leverage point** especially is problematic. This is an observation that is distant from the centre point of the regression line, $(\bar{x}, \bar{y})$, both vertically along the Y axis and horizontally along the X axis. The name arises because of the levering effect it has upon the regression line, causing it to be mispositioned with respect to the rest of the data.

7.10   What to do if the residuals do not meet the regression assumptions

In the preceding section we identified three problems with the residuals of our model of average property price. First, there was an extreme outlier (and possible leverage point) – observation 16. Secondly, the variance of the residuals seemed to be related to the fitted Y values. Thirdly, the residuals show signs of positive spatial autocorrelation.

The pragmatic solution for the outlier is to drop it from the data and refit the model without it. An alternative strategy is to fit a model of the form

$$y = \beta_0 + \beta_1 x_1 + \beta_2 x_2 + \varepsilon \tag{7.28}$$

X_2 is a '**dummy variable**', a set of values with as many observations as the other variables in the model and all but one of its values set to zero. That exception is the 16th value, which will be set to equal one.

Dummy variables are used to separate different categories of data within the model. In modelling responses to a social attitudes survey we could look at the difference between male and female respondents by creating a dummy variable that assigns all males a value of one and all females a zero. The model would then allow us to see whether the mean response of males was greater or less than for females, and whether significantly so. In the same way, by giving an unusual observation its own dummy variable we are examining whether it is categorically different from the rest.

Table 7.3 summarises the results of the model. The regression coefficient for the dummy variable (β_2) is indicating the average house price is £167 238 greater in neighbourhood 16 than the regression model would otherwise predict. This is a significant difference at a greater than 99.9% confidence ($p < 0.001$).

The second problem potentially is harder to deal with. It is one of **heteroscedasticity**, of non-constant variance. This is revealed by the residuals appearing to funnel outwards from the zero line in Figure 7.8(d) (they could also funnel inwards: see Figure 7.13). Heteroscedasticity does not bias the model (the estimates of β_0 and β_1 should be accurate) but it does raise the standard error and therefore the p values.

A technical solution for heteroscedasticity is to fit the regression model using weighted least squares, decreasing the influence of observations that deviate most from the regression line. However, that leaves unanswered the issue of why the heteroscedasticity has occurred. Heteroscedasticity can indicate that one or more of the variables need transforming, that other predictor variables need to be included in the model or that the analysis could benefit from a more explicitly geographical form of analysis (Chapter 9).

To test if there is a relationship between the residual variation and the predicted y values, a second regression line can be fitted where the predicted values are the independent variable and the residual values, squared, are the dependent variable. If this line and its associated p value suggest a statistically significant relationship then there is

Table 7.3 Summary statistics for a regression analysis of the relationship between mean property prices and the percentage of the population with higher educational qualifications in Bristol, with a dummy variable included for observation 16

	Estimate	Standard error	t	p
β_0	108 370	8445	12.8	<0.001
β_1(higher education)	3302	298	11.1	<0.001
β_2(dummy variable)	167 238	27 529	6.08	<0.001
R^2	0.848			
$F_{(2,32)}$	89			<0.001

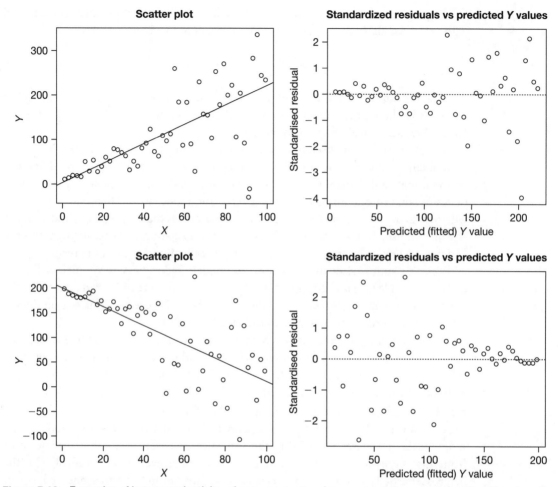

Figure 7.13 Examples of heteroscedasticity, of non-constant variance around the regression line. Note how the observations funnel in or out

a problem of heteroscedasticity in the original model. The residuals are not independently and identically distributed. Instead, the (squared) errors increase or decrease in relation to the predicted *y* values. The logic of this analysis is the basis of the **Breusch–Pagan** test for heteroscedasticity. Fortunately the test reveals that the problem of heteroscedasticity is not as great in our model of property prices as first suggested.

Finally, there is the issue of **spatial dependency** to consider with the residuals showing a geographical patterning. The technical concern is that we have, in effect, less than *n* independent observations and so the degrees of freedom ought to be lowered, hence the standard errors raised, the *t* values lowered and the *p* values raised. This could mean that a relationship appearing significant at a given level of confidence is not actually so once the residuals' lack of independence is taken into account.

In any case, the geographically minded would want to ask what causes the patterning! It would be useful to know, for example, something more about the housing stock in each area, or crime rates, or average length of residence, or some categorical measure of each neighbourhood's reputation/prestige. The spatial patterning of the residuals may indicate the model is too simple and requires more explanatory variables or that a more geographical approach is required (Chapter 9).

7.11 Multiple regression

In the preceding discussion we alluded to using more than one independent variable to help explain the dependent variable. Imagine we are interested in modelling the equilibrium line altitude (ELA) of a glacier in regard to total precipitation accumulating during the winter months (measured in mm) and also the average temperature during the summer months (measured in °C). A model can be fitted of the form

$$y = \beta_0 + \beta_1 x_1 + \beta_2 x_2 + \varepsilon \tag{7.29}$$

which appears the same as in Equation 7.28, except X_2 is here a second continuous variable, not a dummy variable. The ELA is the snow line, the altitude above which the snow does not melt.

The idea of **multiple regression** is to separate out the effects of each of the independent variables upon the dependent variable. In Equation 7.29 there are two independent variables (the two X) but there could be more.

The results of the model are shown in Table 7.4. The amount of precipitation lowers the snow line whereas the summer temperature appears to raise it. Whilst only the precipitation variable is significant at a conventional level of confidence (for example, 95% confidence; $p < 0.05$), we note that the sample size is small ($n = 19$). Referring back to our discussion of statistical power (Chapter 6, Section 6.4) we might be willing to accept a 90% confidence level (which the temperature variable passes), at least until we have more data.

The line of best fit is now really a plane (see Figure 7.14) but without worrying about the terminology we can say it has the equation

$$\hat{y} = 1601 - 0.355 x_1 + 46.7 x_2 \tag{7.30}$$

It suggests that a 1 unit increase in precipitation decreases the ELA by 0.355 units and (if the relationship is regarded as significant) that a 1 unit increase in temperature increases the ELA by 46.7 units.

Table 7.4 Summary statistics for a multiple regression analysis of equilibrium line altitude (Y) against winter precipitation (X_1) and summer temperature (X_2)

	Estimate	Standard error	t	p
β_0	1601	348	4.60	<0.001
β_1 (precipitation)	−0.355	0.064	−5.52	<0.001
β_2 (temperature)	46.7	26.6	6.08	0.099
R^2	0.679		Adjusted R^2	0.639
$F_{(2,16)}$	16.9			<0.001

Source of data: Professor L. Hepple.

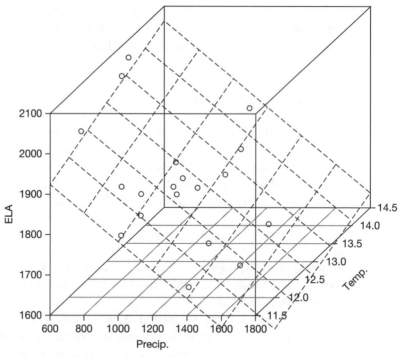

Figure 7.14   The 'line' of best fit for the multiple regression model

7.12   Interpreting multiple regression

Goodness-of-fit

Table 7.4 includes a new goodness-of-fit measure: the **adjusted R^2**. This modifies the R^2 value to make allowance for the number of predictor variables. The logic is that the more X variables there are, the more they will explain the variation in the Y variable.

The adjusted R^2 only increases if each additional X variable improves the model by more than is expected by chance alone.

The adjusted R^2 value is always less than or equal to the original R^2 value and can be used to compare basically similar models of the same data set where only slight changes to the number of predictor variables or to the number of observations in the model have been made. However, it should not be used for comparing different studies where the data and the models are different. Better comparative measures include the **Akaike Information Criterion (AIC)** and the **Bayesian Information Criterion (BIC)**, both of which consider the amount of residual error with respect to the number of observations and the number of predictor variables in the model. The lower the AIC and BIC values are, the better the predictive ability of the model.

Key concept 7.9   Goodness-of-fit measures

The **R^2** and **adjusted R^2** values are goodness-of-fit measures considering how much of the variation in the Y variable is explained by the regression model. The R^2 value ranges from 0 to 1 whereas the adjusted R^2 value, which is better for multiple regression, is always less than the R^2 value and can, occasionally, be negative.

The **Akaike Information Criterion (AIC)** and the **Bayesian Information Criterion (BIC)** are also goodness-of-fit measures. For these, the lower the value, the better the fit.

Standardising the variables

Precipitation and temperature have different measurement units (mm and °C, respectively) so their effects on ELA are not directly comparable. Because of this, it can be advantageous to standardise the data by separately converting all the variables (including the Y variable) to z units (see Equation 7.2).

The results having done so are shown in Table 7.5. The t and p values for β_1 and β_2 are unchanged, as are the R^2 and F values. The y intercept becomes zero.

Table 7.5   Summary statistics for a multiple regression analysis of equilibrium line altitude (Y) against winter precipitation (X_1) and summer temperature (X_2) where each of the variables has been standardised (converted into z values)

	Estimate	Standard error	t	p
β_0	0.000	0.138	0.000	1.000
β_1 (precipitation)	−0.078	0.142	−5.519	<0.001
β_2 (temperature)	0.025	0.142	1.751	0.099
R^2	0.679		Adjusted R^2	0.639
$F_{(2,16)}$	16.9			<0.001

Because they are now measured in standardised units the estimates of β_1 and β_2 can be compared directly. We see that precipitation has approximately three times greater effect on ELA than temperature (compare -0.078 with 0.025).

7.13 Assumptions of multiple regression

The assumptions of multiple regression are the same as for simple, bivariate regression (linearity, the X explain the Y, the residuals are i.i.d.) and the same checks of the residuals should be made (checking for normality, randomness, no heteroscedasticity and for adverse effects of any outliers).

In addition, checks should be made for **multicollinearity**, a problem of the X variables being so related to each other that their separate effects upon the Y variable no longer are reliably and accurately estimated. This can lead to the erroneous conclusion that none of the independent variables relates significantly to the dependent variable.

There is no multicollinearity when the X variables entirely are independent and therefore not correlated with each other. This is the ideal. At the other extreme, there is complete multicollinearity when one or more of the X variables is perfectly correlated with another. This does at least have the virtue of being easily detectable: either an error will be generated or the statistical software will reject the repeated variables. The problem is the more common middle ground.

An indication of multicollinearity is given by calculating Pearson's correlation, r, pairwise for each of the X variables. If the correlation between any two X variables is relatively high (say, $r > 0.5$ or $r < -0.5$) it may suggest a problem. It is not, however, a reliable test. This is because, if the X variables are correlated, then the correlation coefficient for any two X variables will be affected and distorted by other correlations between the Xs.

A better indicator of multicollinearity is the **variance inflation factor** (vif score), interpreted as an index of how much each one variable is correlated with the others in a regression model. As a guide, a vif score of 5 or above suggests multicollinearity, and a score of 10 or above suggests severe multicollinearity. An absence of multicollinearity is indicated by a score of 1.

There are a number of strategies that can be adopted to deal with multicollinearity. The most pragmatic is to drop from the model one or more of the variables exhibiting the problem. A second is to consider whether the variables are correlated because they have a joint effect upon the Y variable. Consider a model of earnings predicted by age and gender (amongst other variables). It is generally the case that male earnings are higher but perhaps that differential increases or decreases with age. If it does we might also consider the joint effects, most simply modelled by multiplying the two separate variables together to create an **interaction term**, as in the following model:

$$y = \beta_0 + \beta_1 x_1 + \beta_2 x_2 + \beta_3 (x_2 \times x_3) + \ldots \qquad (7.31)$$

A third is to consider that the reason why the variables are correlated is because they each touch upon a smaller number of underlying factors that are themselves the key to explaining the Y variable. The idea is then to reduce the variables to those factors. Methods to do so include **factor analysis** and **principal components analysis** (see Robinson 1998).

Key concept 7.10   Assumptions of OLS regression

Ordinary least squares (OLS) regression provides the best linear unbiased estimates of the unknown values β_0, β_1, etc., provided certain assumptions are met.

In addition to linearity, OLS assumes the residuals are independent and identically distributed. This means they should appear normal with no evidence of heteroscedasticity (non-constant variance) around the regression line. There should be no severe outliers and no patterning of the residuals when, for example, the residuals are mapped. The X variables should explain the Y, not the other way around or with feedback loops.

In addition, multiple regression requires checks for multicollinearity between the X variables.

7.14 Partial regression plots

In the presence of multicollinearity ordinary scatter plots of the relationship between Y and each separate X can become deceptive because of the effects of the other X upon them. Partial regression plots are used first to take out the effects of the other X variables leaving a better idea of the actual relationship between the Y and an X (Dunn 1989). They help to reveal whether the data need transforming and whether there are any immediate problems such as outliers.

Imagine we want to form a multiple regression model with a Y variable and four X variables (X_1, X_2, X_3, X_4) but want first to explore the relationship between the Y and each of the X variables individually, to gain a better understanding of the data. Beginning with the relationship between Y and X_1, a partial regression plot is produced by the following stages:

- Fit a regression model with Y as the dependent variable and X_2, X_3 and X_4 as the independent variables. The residuals from this model are what is left in Y once the effects of X_2, X_3 and X_4 are taken out. 'What is left' includes the effects of the variable of interest, X_1.

- Fit a regression model with X_1 as the dependent variable and X_2, X_3 and X_4 as the independent variables. The residuals from this model are what is left in X_1 once the effects of X_2, X_3 and X_4 are taken out.

- Produce a scatter plot with the first set of residuals on the vertical axis and the second set of residuals on the horizontal axis. This gives an indication of the relationship between Y and X_1 having controlled for the effects of the other variables. Note any outliers and/or whether the variables need transforming.

- Repeat the process for each of the other X variables in turn. For example, for X_2, regress Y against X_1, X_3 and X_4; regress X_2 against the same; calculate the two sets of residuals; and plot them on a scatter plot.

7.15 A strategy for multiple regression

In principle a multiple regression model can contain as many X variables as the analyst can acquire from the data sources available. In the era of online data archives and ease of access to large secondary data sets this is a temptation. However, it is also the scattergun approach to modelling and has at least three drawbacks. First, it suggests a lack of engagement with the wider research literature to help select the variables most relevant to the study. Secondly, given enough variables, it is inevitable that some will appear to have a significant effect on the Y variable – if there are 100 independent variables then, at a 95% confidence level, five are expected to show a significant effect on Y due to chance alone. It also is inevitable that some will exhibit multicollinearity. Thirdly, the resulting model may lack a sense of real-world meaning. It may be very difficult to interpret.

A better strategy is to form an initial variable selection informed by theory and/or by previous studies. Having done so, a top-down strategy can be employed beginning with an initial exploration of the data, using simple descriptive statistics and graphical techniques including bar, quantile and partial regression plots to check for linearity, skew, potential outliers, and so forth, and then proceeding to discard those variables that are insignificant predictors of the Y variable and/or outliers, including leverage points that distort the regression analysis.

The top-down strategy is not intended as a one-shot process whereby all insignificant X variables and all outliers are removed in a single sweep. Instead the process will be iterative – taking out a variable or an outlier, seeing what happens and, if necessary, putting it back in the model if removing it did not improve the model.

An improvement will not be judged by a single metric. All the regression diagnostics will be considered in the round – each variable's t and p values, the adjusted R^2 value, the distribution of the residuals and whether they meet the assumptions of regression.

When omitting outliers (or using dummy variables), knowing when to stop is important. There will always be some observations further from the regression line than others. But are they actually having a detrimental effect upon the regression line? To answer this, looking at the standardised residuals will be helpful.

As a guide, **parsimony** should be sought. Parsimony means that the simplest model, the one with the least independent variables, is to be preferred in the absence of a compelling reason for the model to be more complicated. A good test of a model is whether the results can be interpreted and are meaningful in the real world.

Often there is no one right model but there are better and worse ones. A good model is one that is rooted in theory, can be explained, has been developed carefully and where the analyst has used the various statistical tools and information available to inform their decision-making process without resorting to mechanical thinking such as 'the model with the highest R^2 value is best' (which is not necessarily so).

> **Key concept 7.11   Occam's Razor and the minimal adequate model**
>
> **Occam's Razor** is the idea that, when faced with competing and equally good explanations for something, the simplest is to be preferred. In statistical modelling this principle becomes one of **parsimony**: of avoiding overly complex models wherever possible and preferring those that lend support for simpler explanation and interpretation.
>
> Crawley (2005; 2007) describes the aim of statistical modelling as finding a **minimal adequate model**. The process involves going from a 'maximal model' containing all the variables of interest to a simpler model that fits the data almost as well by deleting the least significant variables one at a time (and checking the impact on the model at each stage of doing so). As part of the process, consideration also needs to be given to outliers and to other checks that the regression assumptions are being met.

7.16 The strength of the effects and the problem of 'too much power'

Earlier we alluded to the problem of having too little data to say with reasonable confidence that an X variable is a significant predictor of Y. With little data to go by it is hard to determine if the variables are related, therefore the null hypothesis of no effect will tend to be kept, at the risk of committing a Type II error (not accepting the alternative hypothesis when it is, in fact, true: see Chapter 6, Section 6.4). This is a case of the statistical test having too little power to detect an effect because of the small sample size. This is an issue that students who collect (insufficient) data in the field will need to consider.

However, users of large secondary data sets may find the opposite problem – that every X variable appears to be a significant predictor of the Y variable. This situation arises when there is so much data in the regression analysis that effects are almost bound to be detected even if they are really quite negligible or have little real-world meaning.

One way to address the problem is to take a random and smaller sample of the data and analyse that instead. It is not a very satisfactory solution, though, as it amounts to throwing away data. A second strategy is to get away from the idea of fitting a single model to the whole data set and see, for example, if there are differences in the effects and their statistical significance in different parts of the study region. This is the approach Harris *et al.* (2010) take and is one we look at more closely in Chapter 9. It can, however, be computationally intensive.

A more general strategy is to look carefully at the strength of each X variable's effect on Y as measured in the data. That is, look at the estimates of β_1, β_2, etc., each measuring how much the Y variable is expected to change given a 1 unit change in an X. Do some variables have much stronger effects than others? Are others essentially negligible? The comparison is made easier if the variables are standardised (Section 7.12, above).

Judgements can also be made by whether the omission of a variable has any substantial impact on the overall fit of the model. To gauge this, goodness-of-fit measures are used (adjusted R^2, AIC, BIC), as are F tests (see below).

7.17 Using the *F* test to compare regression models

An ANOVA can be used to help choose between competing regression models. Imagine, for example, two models are fitted using a top-down strategy. The first, model A, has three independent variables (X_1, X_2 and X_3) and the second, model B, has just one (X_1 and X_2). You are interested in checking whether the model simplification was justified.

An F test can be formed to see whether the relative decrease in the sum of the residual error squared (SSE, see Equation 7.23) is greater than the relative decrease in the degrees of freedom that results from removing the variable from the model:

$$F_{(df_A - df_B, df_C)} = \frac{(SSE_A - SSE_B)/SSE_A}{(df_A - df_B)/(df_A)} \tag{7.32}$$

where the degrees of freedom are equal to $(n - p - 1)$, n is the number of observations and p is the number of X variables in each model.

The information required to calculate the F test is set out in the form of an ANOVA table. Table 7.6 is an example where average neighbourhood property price (see Section 7.1) has been modelled, first with four variables (percentage of the population with higher education qualifications, percentage who cycle to work, a measure of crime rate and a measure of average weekly income) and then with three (omitting the cycling variable). In this case the model simplification seems justified because it does not lead to a significant reduction in the explanatory power of the model – the F statistic is insignificant at a 95% confidence level, for example. However, if we go one step further and try removing the income variable we find that simplification is not justified: the F statistic is significant at a greater than 99.9% confidence level – see Table 7.7.

Table 7.6 ANOVA table comparing two models. In this example the model simplification appears justified

Model	df	SSE	$df_B - df_C$	$SSE_B - SSE_C$	F	p (>F)
A	30	13 823 565 843				
B	31	14 523 408 457	−1	−699 842 615	1.519	0.227

Table 7.7 ANOVA table. In this example the further model simplification appears not to be justified (the F statistic is too great)

Model	df	SSE	$df_B - df_C$	$SSE_B - SSE_C$	F	p (>F)
B	31	1.452×10^{10}				
C	32	4.506×10^{10}	−1	-3.054×10^{10}	65.18	<0.001

7.18 Uses of regression

Regression models can be used for different purposes: to summarise data, to make predictions (what would be the value of y if x is . . .?) and to explain what causes what.

The third of these purposes, Freedman (2009, p.1) notes, 'is the most slippery'. **Causal inferences** are most defensible when we have the sort of randomised controlled experiment outlined at the beginning of Chapter 5 as an agricultural trial. If one group (of land, people, etc.) is exposed to some treatment, another is not, and the only remaining differences between the groups can be taken to be random, then it is relatively straightforward to determine whether the treatment does or does not have an effect.

However, many environmental and socio-economic studies are not based on randomised controlled experiments but are more opportunistic, making use of data collected for reasons other than the study itself or taking measurements after the event, hoping to reveal differences between groups that already have been created. These are **observational studies**.

The trouble with observational studies is that although they may reveal associations, determining causes is complicated by other (unmeasured) factors. For example, Freedman (2009), records how cross-national comparisons have revealed a strong correlation between telephone lines per capita and the death rate from breast cancer. This is not because talking on landlines causes cancer but more probably because women in richer countries have fewer children and pregnancy – especially early first pregnancy – is protective.

The relationship between telephone lines and breast cancer rates is confounded by other factors. To some extent multiple regression controls for this by allowing various explanatory variables to be included in the model and aiming to separate the strength and significance of their various effects. However, all observational studies must be treated with some caution, giving consideration to how the data have been collected and what biases they may contain.

Careful research design minimises the problem of confounding and permits more robust conclusions to be drawn. On occasions the researcher may benefit from a **natural experiment**. That is, when one group has been exposed to something whilst another has not, yet in all other ways the groups are extremely similar (and are probably unaware of their exposure).

Freedman gives the example of John Snow's study of the cholera epidemic in London during 1853–1854. Two water companies, Southwark and Vauxhall Company and Lambeth Company, competed and ran separate pipes over an extensive part of London, both supplying a mix of houses and population. In 1852 the Lambeth Company moved its intake pipe upstream into purer water. By showing the death rate from cholera was about nine times greater for houses supplied by the Southwark and Vauxhall Company than by the Lambeth Company, Snow obtained persuasive evidence that cholera is caused by a waterborne virus. Another important feature of Snow's work was how he then used maps to help promote (and dramatise) his research findings (Shaw *et al.* 2001).

7.19 Other types of regression model

Throughout this chapter we have focused on OLS regression, used for continuous data where the regression residuals are normally distributed. For other types of data the assumption of normality will not apply.

If the regression residuals (the errors) are strongly skewed, have a flat or peaked distribution, should be strictly bounded within a (limited) range of values or have nonsensical negative values, then **generalised linear models** (GLMs) may be used instead of OLS. Detailed discussion of GLMs is beyond the scope of this book but it is useful for you to know that they allow different error distributions to be specified that are not normal. For example, Poisson errors for count data and binomial errors for binary data or for data on proportions. For further information see Pampel (2000) and Crawley (2007, Ch. 13).

A further limitation of OLS is that it is sensitive to outliers as we have seen. **Robust regression** offers what Maindonald and Braun (2006, p.153) describe as 'a half-way house between including outliers and omitting them entirely'. It does so by down-weighting them. The need to do so arises because traditionally regression models (like many other statistics) focus on the mean; they 'summarise the relationship between the response variable and predictor variables by describing the mean of the response for each fixed value of the predictors' (Hao and Naiman 2007). However, we could also use the median, any other quantile or, indeed, more than one quantile at once. To do so is the basis of quantile regression.

Finally, in the presence of errors revealing a geographical patterning, methods need to be adopted that take into account and do not ignore the geography. We turn to such methods in Chapter 9, first looking at ways to reveal and quantify patterns of spatial association in the next chapter.

Key points

- A scatter plot is used to explore the relationship between two variables.

- The degree of association between two variables is measured using a correlation coefficient.

- Pearson's correlation coefficient, r, is used for two continuous random variables that are linearly related and normally distributed. It ranges from −1 (a perfect negative relationship) through 0 (no relationship) to +1 (a perfect positive relationship).

- Regression fits a line of best fit to some data, aiming to explain the variation found in the dependent (Y) variable with respect to one or more independent (X) variables. The model takes the form $y = \beta_0 + \beta_1 f_1 + \beta_2 f_2 + \ldots + \beta_k f_k + \Sigma$ where k is the number of X variables.

- The beta values are the regression coefficients. Of these, β_0 predicts the value of y when $x = 0$. The others (β_1, β_2, etc.) are the slopes and indicate the effect of each X variable upon the Y variable.

- An effect that is observed in the data could be due to chance. A hypothesis test is undertaken with a null hypothesis of zero effect for each of the X variables. If the null hypothesis can be rejected at a suitable level of confidence, there is a statistically significant relationship between that X and the Y.

- However, a good model should have more than statistical significance. It should also have real-world meaning so the results can be interpreted and explained. In general, simpler models are preferred.

- The difference between the y value found in the data and what the model predicts given its x value is a residual error. Ordinary least squares (OLS) regression minimises the sum of the squared residuals across all observations in the model.

- OLS regression provides the best linear unbiased estimates of the regression coefficients provided certain assumptions are met. These include the residuals being independent and identically distributed: they are normal, have no heteroscedasticity, are unrelated to the X and Y variables, and have a random patterning on a map.

- It is important also to check for outliers, especially leverage points; and for multicollinearity amongst the X variables when fitting a multiple regression model.

- A top-down strategy for multiple regression aims to find the minimal adequate model – a parsimonious model that fits the data well. Decisions about which variables and observations to omit from the model need to be taken in the round using a range of metrics including goodness-of-fit measures (for example, adjusted R^2), F tests, t and p values of the independent variables and plots of the regression residuals.

- Observational studies limit the ability to say X causes Y. Multiple regression helps by controlling for other factors. Whilst statistical association does provide circumstantial evidence for causation, there is always the possibility that the true cause has not been found.

References

Bivand, R.S., Pebesma, E.J. and Gomez-Rubio, V. (2008) *Applied Spatial Data Analysis with R*, New York: Springer.

Bostock, L. and Chandler, F.S. (2000) *Core Maths for Advanced Level*, 3rd edn., Cheltenham: Nelson Thornes.

Crawley, M.J. (2005) *Statistics: An Introduction using R*, Chichester: Wiley.

Crawley, M.J. (2007) *The R Book*, Chichester: Wiley.

Dunn, R. (1989) Building regression models: the importance of graphics. *Journal of Geography in Higher Education*, 13(1), 15–30.

Freedman, D.A. (2009) *Statistical Models: Theory and Practice*, 2nd edn, Cambridge: Cambridge University Press.

Hammond, R. and McCullagh, P.S. (1978) *Quantitative Techniques in Geography: An Introduction*, 2nd edn, Oxford: Oxford University Press.

Hao, L. and Naiman, D.Q. (2007) *Quantile Regression, vol.* 149 (Quantitative Application in the Social Sciences), illustrated edn, Thousand Oaks, CA: Sage.

Harris, R., Singleton, A., Grose, D., Brunsdon, C. and Longley, P. (2010) Grid-enabling geographically weighted regression: a case study of participation in higher education in England. *Transactions in GIS,* 14(1), 43–61.

Hepple, L.W. (2001) Multiple regression and spatial policy analysis: George Udny Yule and the origins of statistical social science. *Environment and Planning D: Society and Space,* 19(4), 385–407.

Maindonald, J. and Braun, J. (2006) *Data Analysis and Graphics Using R: An Example-based Approach,* 2nd edn, Cambridge: Cambridge University Press.

Monmonier, M.S. (1996) *How to Lie with Maps,* 2nd ed., Chicago: Chicago University Press.

O'Sullivan, D. and Unwin, D.J. (2002) *Geographic Information Analysis,* Hoboken, NJ: Wiley.

Pampel, D.F.C. (2000) *Logistic Regression: A Primer,* Thousand Oaks, CA: Sage.

Rice, M. and Harris, G. (2005) Comparing effect sizes in follow-up studies: ROC area, Cohen's d, and r. *Law and Human Behavior,* 29(5), 615–620.

Robinson, G.M. (1998) *Techniques and Methods in Human Geography,* Chichester: Wiley.

Rogerson, P.A. (2006) *Statistical Methods for Geography: A Student's Guide,* 2nd edn, London: Sage.

Shaw, D.M., Dorling, P.D. and Mitchell, D.R. (2001) *Health, Place and Society,* Harlow: Prentice Hall.

Wilkinson, R. and Pickett, K. (2009) *The Spirit Level: Why More Equal Societies Almost Always Do Better,* London: Allen Lane.

Detecting and managing spatial dependency

Chapter overview

A key theme of this chapter is to gain an understanding of how spatial dependency may distort statistics, and how to measure the potential effect.

In this book, we have focused on introducing you to relatively straightforward statistical methods; the methods introduced so far assume that there is no spatial autocorrelation in your data. The reality of the situation is likely to be far from this in many cases, and the advanced analyst will either make special adjustments to standard formulae that compensate for spatial dependency, or seek out alternative statistical tests that explicitly account for spatial autocorrelation. Understanding these adjustments is not expected for an introductory course in statistics but what is important is that you appreciate the potential importance of this fundamental geographical problem and can take steps to assess associated issues. We outline in particular the modifiable areal unit problem and the ecological fallacy.

Being able to measure spatial pattern, and to assess how this varies at different scales, has other benefits apart from aiding the process of statistical testing. It is often an initial first step towards investigating more about geographical processes, be they in the realm of human or physical geography. This chapter introduces both global and local methods of assessing spatial dependency (autocorrelation), including the variogram, Moran's I and Getis's G* statistics.

Learning objectives

By the end of this chapter you will be able to:

- Understand why spatial dependency is an issue in geographical analysis.
- Describe and exemplify the ecological fallacy;
- Define the modifiable area unit problem and exemplify it with numerical examples;
- Reflect on the effect of spatial pattern on the statistics you have learned to this point in the book.
- Identify and measure spatial dependency (spatial autocorrelation) in different types of data, in particular:
- Distinguish between local and global measures of spatial association;
- Define and interpret a semi-variogram as a measure of global autocorrelation;
- Define and interpret measures of global autocorrelation in gridded (raster) data;
- Define and interpret measures of local autocorrelation in gridded (raster) data.

8.1 Introduction

We first introduced the concept of **spatial autocorrelation** Tobler's first law (Key concept 1.9) to you back in Chapter 1 (Section 1.6). As this is key to the issue of spatial dependency, the particular focus of this later chapter, let us recap again here.

Key concept 8.1 More about spatial autocorrelation

Spatial data have a tendency to be more similar in value at nearby locations than those further away, a phenomenon called **spatial autocorrelation**.

Global autocorrelation is the overall degree of spatial association within an entire data set. **Local autocorrelation** is the degree of spatial association in data surrounding a particular location within a certain radius or area.

Positive spatial autocorrelation means that if you take measurements of some geographical feature or phenomenon, similar values will be found for data from locations that are situated close together. Finding that the values of neighbours are 'opposite' to each other is evidence of **negative spatial autocorrelation**.

In either case, there is a spatial (geographical) dependency within the data, although it may not be so simple as to say what is measured in one place is caused by what happens in places around it.

Tobler's first law of geography (Key concept 1.5) is a somewhat unintentional rendition of the concept of spatial autocorrelation by the geographer Waldo Tobler (1930–):

> Everything is related to everything else, but near things are more related than distant things.

Actually, this is a statement of some controversy, not least because Tobler himself did not mean it to be taken as a universal 'law' and because it might be claimed that it denies the essence and variation of human individuality, particularly in our multilingual, multi-ethnic and highly digitally connected world. However, the 'death of geography' has been greatly exaggerated! Interactions and effects due to proximity remain important.

Miller (2004) draws together some accessible and considered reflections on whether Tobler's first law (TFL) remains a useful concept in our increasingly fragmented (Couclelis and Getis 2000) cyber-linked environments. He concludes that if we also take the temporal dimension of connectivity into account, and consider the fact that complexity often arises from simple local interactions, TFL *does* continue to be a core of spatial analysis and quantitative geography.

This chapter starts by considering in more detail why spatial dependency matters when conducting quantitative analyses of geographical data, focusing on two issues known as the **modifiable areal unit problem** and the **ecological fallacy**. Secondly, the chapter outlines how we can assess **global and local spatial autocorrelation** (Key concept 8.1) in a data set, and hence the degree to which your analyses may be adversely affected by spatial dependency. More positively, we also look at what analysing spatial autocorrelation can reveal in regard to the scale at which processes are manifesting in the human and physical landscape.

8.2 Why does spatial dependency matter in the context of quantitative geographical analyses?

We have noted previously (Chapter 4, Section 4.5) that an assumption of inferential statistics is that the data are independent; a measure of spatial autocorrelation in a data set inherently violates this assumption. Non-independent residuals cause the underestimation of the sum of squares calculation, distorting correlation and regression figures and inflating the value of hypothesis test statistics such as ANOVA (for example, Clifford *et al.* 1989; Diniz-Filho *et al.* 2003). In turn, this potentially gives rise to the incorrect rejection of a null hypothesis (a Type I error). For Pearson's product moment correlation and multiple regression, the greater the spatial autocorrelation between both dependent and independent variables, the greater the error (Lennon 2000). In particular, spatial autocorrelation causes two scale-related phenomena, known as the modifiable areal unit problem and the ecological fallacy. This section focuses on these last two matters.

The modifiable areal unit problem

Within geographical information science (see, for example, Heywood *et al.* 2006; Longley *et al.* 2011) attention is given to the modifiable areal unit problem (MAUP) (Openshaw 1984). The root of the problem is that geographical data are often not available for the specific entities, objects or processes we are interested in studying. Instead, the data have those geographical features grouped with others into zones or areal units. The specific detail is then lost. Consequently, what we discover by analysis is a function not just of the feature's own characteristics but also of the grouping.

Figure 8.1 demonstrates the problem. Imagine there has been a spate of burglaries at the intersection of two roads that define the boundaries of four neighbourhoods. The location of each incident is shown in Figure 8.1(a). Further imagine that the detailed information is not available to the analyst. Instead, for reasons of privacy as well as convenience, the 'raw' government data are grouped into administrative zones. A total count is then reported for each.

Figure 8.1(b) has the resulting count per square unit. In this case the zoning scheme has quartered the data, smoothing out the cluster of burglaries. Had the zones been defined differently, as in Figure 8.1(c), then we would have got a different impression of the pattern of burglary.

Together, Figures 8.1(b) and 8.1(c) show the zoning effect. The impression we get of an area is dependent on the data we assign to it, and that in turn is dependent on where that area is positioned – where we or others draw the lines. It is also dependent on the size and shape of the area. Changing those creates scale effects, as Figure 8.1(d) illustrates.

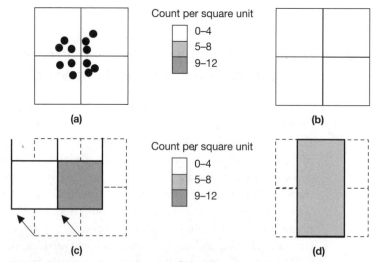

Figure 8.1 The modifiable areal unit problem (MAUP) means that the impression we get of an area is dependent on the data we assign to it: (a) the actual location of burglaries (but not known to the analyst); (b) what the analyst sees – the data aggregated into zones; (c) if the zoning changes then so will our impression of the neighbourhoods; (d) changing the size of the zones leads to scale effects

> **Key concept 8.2 The modifiable areal unit problem**
>
> The **modifiable areal unit problem (MAUP)** arises when the areas of analysis used in geographical research are not natural, definitive or designed with the analysis in mind. More often they were designed by other people, for other reasons such as postal delivery or for governance. The design of the zones – their size and location – will have an impact on the analysis.

An example of the MAUP is in the design of voting areas such as the geographical constituencies used in the UK general elections. Candidates for election become Members of Parliament by winning more of the votes in their constituency than other candidates do. Voters can express only one preference; there are no first or second choices and no transferable votes. It is a winner-takes-all system, for a series of mini-elections held on a constituency-by-constituency basis but all at the same time.

The MAUP arises because places evolve and their population distributions change. This means that constituency boundaries must be reviewed. Since there are few places where natural barriers clearly define a territorial unit, the boundaries of a constituency can usually be questioned, challenged and an alternative offered by those with an interest. A shrewd political party would take the opportunity to try and influence the review process to its favour, to maximise its chances of winning the constituency.

It is the role of the Boundary Commissions to define new constituencies which meet the legal criteria (including balancing the number of electorates per constituency), and they do so without any reference to political data. Moreover, partisan issues cannot be raised in any representations made to them by interested parties. Nevertheless, their non-partisan decision making usually favours one party rather than another because of the MAUP. At the 1997–2005 general elections, Labour was a major beneficiary of this cartographic activity (Gudgin and Taylor 1979; Rossiter *et al.* 1999).

You may have heard of the term **gerrymandering**; this is sometimes the term given to the illegal manipulation of boundaries for political ends that is possible owing to the MAUP. It relates to a newspaper cartoon of the 1812 boundary changes for Essex County, Massachusetts, which were somewhat unfairly attributed to the State Governor Eldridge Gerry (Figure 8.2).

While the classic discussions of the MAUP have focused on human geography, especially in regard to political manoeuvring, it remains critical to consider MAUP effects when undertaking any quantitative spatial analyses. Fotheringham and Wong (1991) go as far as to state that their particular work indicates 'strong evidence of the unreliability of any multivariate analysis undertaken with data from areal units'. What this statement stresses is the importance of checking the geographical units of analysis in a study are fit for purpose.

Analyses in physical geography may consider data by underlying geology or bio-region, for example, both of varying shape and size. Further, the results of many

THE BETTMANN ARCHIVE

GERRYMANDER, a fictional creature based on the shape of an electoral district of Massachusetts, as set up for political reasons.

Figure 8.2   This cartographic cartoon first appeared in the *Boston Centinel* on 26 March
1812, and depicts Essex County (as redistricted by Eldridge Gerry) as a salamander. From this
combination, we get the verb to gerrymander which remains in common usage today

remote sensing (Hay *et al.* 2001) analyses are critically dependent on the scale at which
they are performed. Dark and Bram (2007) provide a useful overview on the subject
of relevance to environmental scientists, while He *et al.* (2007) demonstrate how the
semi-variogram (see Section 8.3 below) can be used to unpack scale and zoning sensitivity within an ecological context. Overall, having an understanding of how the
MAUP may affect your study is key, as is identifying the most appropriate scale for
your analyses.

The ecological fallacy

Related to the MAUP is the ecological fallacy. In general, the ecological fallacy means
that statistical relationships found at one scale of analysis may not hold at other
scales.

Look at Table 8.1. Recalling Pearson's correlation coefficient from Chapter 7 (Key
concept 7.3) it can be seen that, for census zones in England and Wales, there is a
negative relationship between the rate of unemployment and the rate of owning four
or more cars. This implies that areas with more households owning many cars have
less unemployment, which is not a surprising revelation!

Table 8.1 The apparent strength of the relationship, measured for rates of four-vehicle owner-ship and unemployment using Pearson's correlation, r, changes as the scale of analysis does

Scale	n (number of zones)	r (correlation)
Region	9	−0.95
Local authority	376	−0.77
Electoral ward	8868	−0.55

Source: 2001 UK Census.

What is more interesting is how the strength of the relationship varies: at the regional scale the (negative) relationship is near perfect ($r = -0.95$); at the local author-ity scale it is less strong ($r = -0.77$); and at the electoral ward scale (much smaller zones) it is of almost half the strength of the regions ($r = -0.55$).

Why the change? Assume, quite reasonably, that there is a relationship between car ownership and employment; that areas with higher unemployment rates are less likely to have households owning four or more cars. This is a general relationship, not a universal one. Some places will buck the trend.

Further assume that we use the proportion of households owning four or more cars to predict the unemployment rates in each place. Because the relationship is not perfect, in some places unemployment will be over-predicted and in others it will be under-predicted. Now imagine the data are aggregated into larger areas, then larger again, first from electoral wards into local authorities and secondly into regions. If the over- and under-predictions begin to cancel each other out then the strength of the relationship between car ownership and employment will appear to increase with the size of the areal units.

A more specific meaning of ecological fallacy

The more specific meaning is when inappropriate assumptions are made about indi-viduals using grouped data (grouped, for example, by census tract or some other approximation of neighbourhood). This is the original meaning of the term and is associated with, but not used in, a paper by Robinson (1950). In it, Robinson con-sidered the practice of using ecological correlations as a substitute for individual cor-relations. With ecological correlations the statistical object is a group of persons and so 'the variables are percentages, descriptive properties of groups, and not descriptive properties of individuals' themselves (p.351). The substitution rarely is by choice but occurs when individual data are not available. (Individual data may be considered confidential or access prohibited by data protection laws.)

In his paper, Robinson used 1930 US Census data to look first at the relationship between ethnic background and illiteracy, and then at the relationship between being born overseas and illiteracy, both 'considered as properties of individuals' (p.353). The results were then compared with an ecological analysis using percentage data for geo-graphical areas – for nine geographical divisions of the United States and also for states.

Each of Robinson's correlations is shown in Table 8.2. Notice how they change and that some move from being positive to negative. Depending on the scale of analysis, you might consider the foreign born as having higher or lower average literacy in com-parison with the native born.

Table 8.2 William Robinson's demonstration that ecological correlations can be a poor substitute for individual correlation

Scale	Ethnicity vs illiteracy	Immigration vs illiteracy
'Individual'	+0.203	+0.118
Ecological: states	+0.773	−0.526
Ecological: nine census divisions	+0.946	−0.619

Source: See Robinson (1950) for details.

Robinson concludes his paper by stating that:

The relation between ecological and individual correlations which is discussed in this paper provides a definite answer as to whether ecological correlations can validly be used as substitutes for individual correlations. They cannot.

(p. 357)

That conclusion is too strong. The real problem is that the substitution works better in some cases than in others and not always in wholly predictable ways.

However, Robinson was certainly right to draw attention to the problems aggregate data bring when trying to establish causes. Robinson expected those born outside of the United States to have less literacy than those born in it, but the expected relationship was not evident when using the data for states or for the census divisions. An explanation for the anomaly is that although the immigrants did have lower average literacy, they were attracted to areas where the literacy rate amongst the native population was above average. As the foreign-born people constitute only a minority of the population in each area, the net result is correctly to associate the immigrants with places of above-average literacy but wrongly to deduce that the foreign born have higher literacy.

In the Robinson example, a way forward might be to undertake a standard test of reading ability, comparing the scores of foreign-born school pupils with those of native pupils. But even that might not be definitive: what if the two types of pupils attend different schools and some are much better resourced than others? Might the relationship between immigration and literacy be confounded with the effects of financing? And what if 'the immigrants' are actually a diverse group of people with varied social and economic backgrounds? Does the relationship between immigration and literacy vary by some measure of social class, or of gender, or of country of origin? And are the children's experiences representative of their parents' anyway? Perhaps they are sooner 'naturalised' into the new society?

What can you do about avoiding this particular trap? Connolly (2006) highlights the role supplementing summary statistics with further exploratory approaches to avoid

> ### Key concept 8.3 The ecological fallacy
>
> The **ecological fallacy** warns against assuming that a statistical relationship is independent of the scale of analysis. As the scale changes, so might our understanding of the relationship. It can be a mistake to apply what we learn about grouped data to individuals within those groups, and can too easily lead to stereotyping – portraying individuals in inappropriate ways.

unfortunate "labelling". In his case, he identified the need to focus on gender and ethnic difference within grouped statistics on pupil performance so as to avoid 'underachiever' and 'overachiever' types of language. The names used to label neighbourhoods in systems for marketing (geodemographics) can also give rise to stereotyping and misrepresentation of the people who live there.

8.3 Looking for spatial autocorrelation

There has, in recent decades, been a change in the way spatial associations have been measured and the reasons for doing so. Spatial association was originally viewed as a global measure of a data set, a crude trend and a confounding factor when applying traditional statistical measures. This still matters. Quantifying global spatial association can be seen as a statistical 'systems check' to see whether some adapted form of standard inferential statistics is required to tackle your research problem at the particular scale you are working at (for example, Dark 2004).

In more sophisticated usage, however, an awareness of spatial autocorrelation can be used to assist with the design of sampling plans calculated to avoid such statistical compromises. Alternatively, spatial autocorrelation can be valued in its own right to assess multi-scale processes (for example, Weissmann and Fogg 1999) or to build surfaces, in which case a sampling regime that is fit for detecting spatial structures is critical (such as random or systematic-cluster designs). It is towards this latter, more positive, consideration of spatial autocorrelation that we see a more recent turn.

A variety of ways in which we can tap into and assess locally varying patterns of association have been established by researchers such as Luc Anselin (Anselin 1995), Art Getis (Getis and Ord 1992) and Chris Brunsdon (Brunsdon *et al.* 1996) over the past 15 years. These researchers measure spatial association explicitly in order to reveal local variation and pattern in relationships between variables, as opposed to seeing them as something of a statistical nuisance.

As we have highlighted in Section 8.1 spatial association can be measured both globally across an entire data set and locally to particular areas (Key concept 8.1). **Positive spatial autocorrelation** means that if you take measurements of some geographical feature or phenomenon, similar values will be found for data from locations that are situated close together. Finding that the values of neighbours are 'opposite' to each other is evidence of **negative spatial autocorrelation**.

In this section, we expand on how you might go about quantifying these measures of spatial association in a variety of ways. First, however, we briefly need to consider what we mean by distance and neighbourhood, and also consider notions of symmetry in spatial autocorrelation.

Distance and neighbourhood

The way in which you treat the concept of neighbourhood will vary depending on the type of data you are investigating. First, let us consider point data. In this case (Figure 8.3(a)), including the values of data points that are within a set distance from a given location is the simplest solution. When you are working with gridded raster (master), data, however, your data fall more naturally into blocks as opposed to rings; the

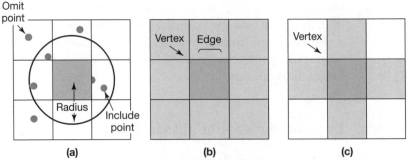

Figure 8.3 (a) Points within a set radius of the centre of a grid square are assigned to the spatial association computation, and those that fall outside the circle are not; (b) queen's case contiguity for gridded data, in which all grid squares with either edges or vertices touching the central square are assigned to the spatial association computation; and (c) rook's case contiguity, in which only grid squares with edges touching the central square are assigned to the spatial association computation

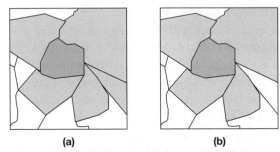

Figure 8.4 (a) Queen's case contiguity for zonal data, in which all zones with either edges or vertices touching the central square are assigned to the spatial association computation. (b) Rook's case contiguity, in which only zones with vertices touching the central square are assigned to the spatial association computation

question arises as to how you should deal with the corner segments of your block. In this case, you have a choice whether to incorporate the corner squares of your block that touch your area either by corner (**vertex**) or **edge** (Figure 8.3(b)), or whether to consider only the elements that match by edge (Figure 8.3(c)). The former of these two situations is known as **queen's case contiguity,** and the latter is known as **rook's case contiguity**.

When working with zones as opposed to points or grids, for example when using census enumeration districts, it is common for zones to be included in the computation of spatial association measures if they are immediate neighbours of the zone being assessed, but are otherwise excluded from the analysis (Figure 8.4).

Symmetry of spatial association

Many factors can influence measures of spatial association. For example, when investigating temperature, latitude introduces a north–south trend. Equally, rainfall is heavily influenced by the prevailing wind patterns. Dominant hydrologies relating to elevation or geological outcrops may influence soil chemistry unevenly. These are just a few potential causes leading to a pronounced directional pattern of spatial association. In your analysis, you can control for the obvious factors, either by removing a trend from

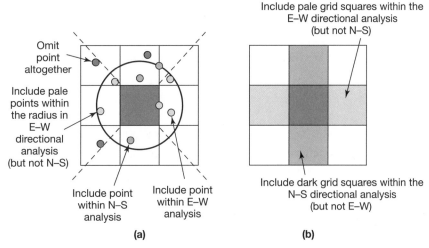

Figure 8.5 Anisotropic approaches to data gathering for the measurement of spatial association, in (a) point data and (b) gridded data

the data set or by measuring spatial association separately in different regions of a larger data set. You may well wish to compute a measure of spatial association in multiple directions, either to confirm the importance of the directional influence of particular environmental circumstances or to explore the characteristics of the data set more fully.

Measures of spatial association that assume equal effects in all directions are known as **isotropic** measures; those that compute spatial autocorrelation in specific directions are known as **anisotropic** measures. Figure 8.5 exemplifies how measurement might be adapted in order to evaluate anisotropy in the point (a) and gridded (b) cases. Note that, unless you are computing spatial association based on binary presence/absence data, spatial association is a function of both data location and value. Just because the point patterns in Figures 8.3 and 8.5 are not symmetrical does not mean that spatial association is uneven.

These concepts of distance and isotropy apply to the measurement of spatial autocorrelation across a variety of quantitative approaches, both local and global in nature.

8.4 Global measures of spatial autocorrelation

The semi-variogram

The semi-variogram is perhaps one of the most classic measures of global spatial association. It is commonly used to explore a data set visually, in order to estimate how far you might need to move away from a particular data point before another data point separated by that distance can be considered as unassociated with it. This information can also be used as input to an **interpolation** or surface modelling method known as **kriging,** which estimates the value of a surface between known data points at locations where no actual measurements have been made. For more details regarding surface construction, see a general GIS textbook such as Burrough and McDonnell (1998 or consult the comprehensive guide to geospatial analysis available at www.spatialanalysisonline.com); kriging in particular is discussed in more mathematical detail in geostatistical textbooks such as Webster and Oliver (2007).

At the heart of the semi-variogram, as the name suggests, is a measure of variance. We introduced you to variance back in Chapter 2 (Section 2.6, equation 2.8); to recap, the variance of a data set is abbreviated by the letter s^2 and is defined by the following formula, where there are n data points in the data set:

$$S^2 = \frac{\sum_{i=1}^{n}\left[\left(x_i - \bar{x}\right)\right]}{n-1} \tag{8.1}$$

Refer to Section 2.6 at this point if you have forgotten how to manage the summation operator (Σ) and compute the sum of the squares (Key concept 2.8); turn back to here only when you have understood the formulaic logic, as you will need this knowledge to follow the rest of the chapter.

In the case of the **semi-variance** (as opposed to variance), this is computed as a function of distance-away from a known point. For each point in the data set, we compute the distance to every other point. These values are then assigned to a particular spacing or **log distance**. Inevitably, the distance between most data pairs will not conform exactly to the set log spacings; we therefore assign a pair to its closest log, within a certain **log tolerance**. Imagine you have 80 points in your data set, and you have computed the distance between each point in turn to all of the others. Now you draw a histogram of these distances that classifies the values into equally divided bands, for example 0–50 m, 51–100 m, 101–150 m, etc. (Figure 8.6). Overall, you have (80–1) × 80/2 unique distance measures; we divide by 2 here because the distance between points a and b (say) is the same as the distance between points b and a. In this example, the lag distance is 50 m and the tolerance is ±25 m; the values for n in Equation 8.2 are 30, 700, 1669 and 800 for the different distance bands or lags respectively and the values for h 25, 75, 125 and 175 m.

The semi-variance of a data set at a distance h from location x, given n data points falling into this distance category and abbreviated by $\gamma^*(h)$, is a value derived from the data as defined by

$$\gamma * (h) = \frac{\sum_{i=1}^{n}\left[y\left(x_i\right) - y\left(x_i + h\right)\right]^2}{2n} \tag{8.2}$$

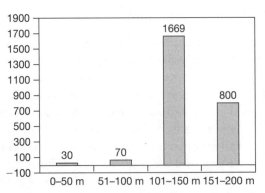

Figure 8.6 Histogram of number of data points falling into particular distance categories (or lag tolerances)

Key concept 8.4   Semi-variance

The **semi-variance** is a measure of how similar points near to one another are, and the rate at which this relationship decays with distance.

The semi-variance of a data set at a distance h from location x, given n data points falling into this distance category and abbreviated by $\gamma^*(h)$, is a value derived from the data and defined as

$$\gamma*(h) = \frac{\sum_{i=1}^{n}\left[y\left(x_i\right) - y\left(x_i + h\right)\right]^2}{2n}$$

Computation of semi-variance implies that there is **stationarity**, or a lack of a directional trend in the data, and that the amount of spatial autocorrelation is constant across the study region.

The **experimental semi-variogram** is simply a plot of semi-variance with distance h, based on known sampled data points. Let's now have a look at a general example of a semi-variogram, with a view to assessing its structural components (Figure 8.7).

Can you see the close connection between equations 8.1 and 8.2? Both use a sum of squares calculation. Also, note that we have different numbers of data points available to us to compute the semi-variance at different lags. Cast your mind back to Chapter 4 regarding sampling; given the variation in data volume, can you be equally confident that you have a good estimate of the semi-variance at every lag distance? A general rule of thumb is that at least 50–100 points are needed to construct a semi-variogram (Burrough and McDonnell 1998, p.137), but this is application dependent and arguably on the low side depending on the internal variation in the number of data points within each log. A further rule of thumb is to avoid computing semi-variance at a lag distance beyond half the width of your study area (Burrough and McDonnell 1998, p.138).

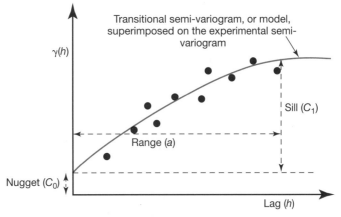

Figure 8.7 The experimental semi-variogram, showing a typical pattern of semi-variance $\gamma(h)$ plotted against log distance. A transitional semi-variogram, or model, is superimposed on the experimental semi-variogram to illustrate the terms commonly used to describe variogram structure (the **nugget, sill** and **range**; see Key concept 8.5)

> ## Key concept 8.5 Semi-variogram
>
> This shows semi-variance (Key concept 8.4) plotted against lag distance.
>
> The **experimental semi-variogram** is a plot of actual semi-variances based on original data against lag distance, where **lag** refers to a distance band within which a set of distance measurements between points fall. The distance measurements used are those from one point to all others in the data set.
>
> The **transitional, or modelled, semi-variogram** is a plot of the best fitted model to the semi-variance data against lag distance. Typical models used for this purpose are the **Gaussian, exponential, linear** and **spherical** models (see Key concept 8.7).

Since the semi-variance is a global measure of association between data at a certain distance, the expected semi-variance should be identical for all observations separated by a particular distance; this applies *regardless of where the data point is located across the area of analysis*. This phenomenon is fundamental to the **regionalised variable theory** which lies behind geostatistical methods such as kriging associated with the semi-variance. This makes checking for anisotropy, by computing a variogram in a variety of directions informed by the wider geographical literature, an important component of the analysis. It also suggests that where pronounced regional variation occurs in the area of your analysis, for example two different geologies or land cover types, you should consider building a variogram per region. The disadvantage of fragmenting your data set in this way goes back to the question of confidence you have in your variogram; you run the risk of computing multiple ill-defined (relatively) 'local' variograms as opposed to a better defined but slightly non-stationary global variogram, depending on context.

Figure 8.7 illustrated the experimental variogram in its most general form. In contrast, Figure 8.8 shows a particular application of an experimental semi-variogram, in this case to assess the spatial dependence in potential insect development (or phenology) across the UK. Insect phenology depends largely on temperature. In the normal case, phenological models are computed in the UK using maximum and minimum temperature data from standard Meteorological Office weather stations. These models, to some degree, build in levels of uncertainty relating to local microclimate. The context for this work is the assessment of risk to British agriculture from a non-indigenous pest or pathogen, should it enter the UK (Jarvis and Baker 2001); most countries apply levels of plant health inspection or import control at their borders depending on the perceived biological threat. In this case, the insect under investigation is the Colorado Beetle (*Leptinotarsa decemlineata*), which is not established in the UK but is resident in nearby areas of Europe.

The semi-variograms illustrated are for potential Colorado Beetle development in the UK, based on estimates from a phenology model using temperature data located at stations of the Meteorological Office recording network, and measured in Julian days of the year (where 1 January is day 1 and 31 December, in a non-leap year, is 365). The lifecycle of the insect goes in the sequence from fully developed adult to egg, larvae, pupae then immature adult. The figure shows that the spatial dependence over the landscape at the larval stage has a longer range than that of the pupa, while for

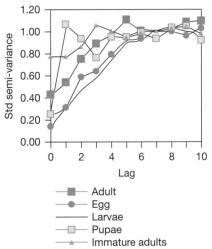

Figure 8.8 Experimental standardised variograms for results of the Colorado Beetle phenology model (Julian dates of emergence), run using observed UK Meteorological Office station data, by stage

immature adults it is not possible to capture the spatial dependence of the insect at the sampling scale made feasible using the UK weather station network. Why might this be the case? The answer lies in the variability of the weather. The longer the insect cycle has continued, the more day-to-day variations in the weather have accumulated at any one place over time; hence, the levels of spatial dependence in the resultant insect phenology are increasingly short range in nature

Importantly for pest risk modelling, the semi-variograms tell us that, *in this particular case,* it would be wiser to assess potential development of the later stages of the insect on estimated temperature surfaces rather than building surfaces based on outputs from phenology models; we simply cannot estimate spatial association, required for surface building, using phenology model outputs. In contrast, while semi-variograms for daily maximum and minimum air temperatures do vary throughout the year, the range is tractable on the majority of days based on the particular sample network available (Jarvis 2000).

Key concept 8.6 Semi-variogram parameters of the exponential model

There are three parameters that define the semi-variogram when using the exponential model:

- The **nugget** (C_0) represents subgrid-scale variation that cannot be estimated by reasons of the sampling grid spacing or because of measurement error.

- The **range** (a) is a measure of the degree of association, or correlation, between data points, represented in terms of distance. It tells us how far we can go from a data point before going beyond the extent of its influence.

- The **sill** (C) is the value of the semi-variance as the lag (h) tends towards infinity; in non-standardised data, it is equal to the total variance of the data set.

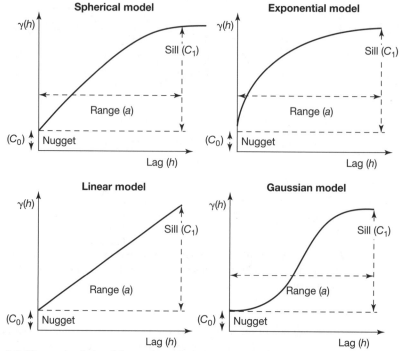

Figure 8.9 The general visual form of semi-variogram models

The experimental variogram is a first step towards assessing variation in a data set in a quantitative way. Having constructed your experimental variogram, where h is the lag distance between pairs of data points, it is also possible to fit a **transitional semi-variogram** (regional mathematical model of the semi-variance) to your data. These models come in a variety of popular forms, but all seek to model the structure of the underlying association between your data points.

Key concept 8.7 Common structures used to model the semi-variogram

Commonly encountered models of the semi-variogram are as follows:
The **exponential** model:

$$\gamma(h) = C_0 + C\left(1 - e^{-h/a}\right) \tag{8.3}$$

The **Gaussian** model:

$$\gamma(h) = C_0 + C\left(1 - e^{-h^2/a^2}\right) \tag{8.4}$$

The **linear** model:

$$\gamma(h) = C_0 + bh \tag{8.5}$$

---

where the slope (b) is the ratio of the sill (c) to the range (a).
The **spherical** model:

$$\gamma(h) = C_0 + C\left(\frac{3h}{2a} - \frac{1h^2}{2a^2}\right) \quad \text{for } h \leq a$$

$$\gamma(h) = C_0 + C \quad\quad\quad\quad\quad \text{for } h > a$$

(8.6)

Common model structures fitted to a semi-variogram are the **linear, exponential, Gaussian** and **spherical** models (Key concept 8.7). Their general shapes are illustrated in Figure 8.9. As you can see, the spherical model approaches the sill more quickly than the exponential model whilst the Gaussian form displays parabolic behaviour near the origin; the other models are linear in form near the origin. In the linear model, the data do not support any evidence for a sill or a range; rather, they show increasing semi-variance as the lag increases. As a measure of spatial association, the linear model is therefore a somewhat frustrating form. Commonly, models are fitted using maximum likelihood or least squares fitting algorithms, but assessment by eye is also important to get a feel for the predominant model structure that is most relevant to your data. If you are interested in exploring these models in more detail, Webster and Oliver (2007) is a good starting place.

8.5 Other measures of global autocorrelation

Today, the use of these global measures is less commonly found within the research literature. However, they are included here both to contextualise discussions and because of their place in geographical history. Cliff and Ord's (1969) work on testing spatial autocorrelation, for example, has been influential in developing statistical methods that explicitly consider scales and types of geographical patterns, particularly in epidemiological settings. Further, as commented above, computing a global statistic can be a first step towards evaluating whether it is appropriate to apply a standard statistical method on a particular geographical data set. We should note, however, that a global statistic indicating no overall spatial association may nevertheless mask local spatial dependency; this special case, as you will be aware by now, would be an indication of the MAUP.

Moran's *I*

Moran's *I* (Moran 1950) is one of the most commonly encountered measures of global association. Computation of the value uses a rook's case neighbourhood (Section 8.1) as follows:

$$I = n \frac{\sum_{i=1}^{n} \sum_{j=1}^{n} w_{ij}(y_i - y)(y_j - y)}{\left(\sum_{i=1}^{n} (y_i - \overline{y})^2\right)\left(\sum_{i=1}^{n} \sum_{j=1}^{n} w_{ij}\right)}$$

(8.7)

where there are n data values, y_i is the value of the variable at grid square i, the global mean of the data set is $\bar{y}$ and the proximity between the locations i and j is supplied by the weight w_{ij}.

The equation is also sometimes expressed as

$$I = n \frac{\sum_{i=1}^{n} \sum_{j=1}^{n} w_{ij} z_i z_j}{\left(\sum_{i=1}^{n} z_i^2\right)\left(\sum_{i=1}^{n} \sum_{j=1}^{n} w_{ij}\right)} \tag{8.8}$$

where z_i are differences in the value of the variable y_i from the global mean $\bar{y}$ (that is, $(y_i - \bar{y})$), and z_j are differences in the values of the variables y_j in the local neighbourhood from the global mean $(y_j - \bar{y})$).

This might appear to be a rather complex formula, but when unpacked it is much easier to understand (see below). In practice, you can readily compute the value of Moran's I in a variety of spatial software or GIS packages; typical input parameters are the size of the neighbourhood window and, in some cases, the particular neighbourhood type you wish to adopt (for example, rook's vs queen's case) and the option to use a bespoke arrangement of weights. You might, for example, wish to give preference to cells nearest the centre of a neighbourhood if computing I for anything larger than a 3×3 grid, perhaps weighing in inverse proportion to distance from the cell in question. Typically, however, weights are set either to 1 or 0 depending on whether they are within a set range or not (see Chapter 9).

In general, values of I range from -1 (absolutely no spatial autocorrelation detectable) to $+1$ (maximal spatial autocorrelation). Random spatial patterns are indicated by a zero score.

Getis's G statistic

Getis's G statistic differs from Moran's I in that it provides a measure of the degree to which high or low values cluster together (de Smith *et al.* 2007, p.228).

Getis's global G statistic is computed as follows:

$$G = n \frac{\sum_{i=1}^{n} \sum_{j=1}^{n} w_{ij}(y_i - y)(y_j - y)}{\left(\sum_{i=1}^{n} \sum_{j=1}^{n} (y_i - y)(y_j - y)\right)}, \quad i \neq j \tag{8.9}$$

or

$$G = n \frac{\sum_{i=1}^{n} \sum_{j=1}^{n} w_{ij} z_i z_j}{\left(\sum_{i=1}^{n} \sum_{j=1}^{n} z_i z_j\right)}, \quad i \neq j \tag{8.10}$$

Unpacking this formula follows a similar logic to that for Moran's I statistic. Similarly to Moran's I, Getis's G can be found as a ready-to-use function in a number of specialist software packages.

Other methods

Getis's G, Moran's I and the semi-variogram are not the only global measures of spatial autocorrelation. Other options that you may come across in GIS software or the

Computing Moran's *I*: Unpacking the formula

$$I = n \frac{\sum_{i=1}^{n} \sum_{j=1}^{n} w_{ij}(y_i - y)(y_j - y)}{\left(\sum_{i=1}^{n} (y_i - \bar{y})^2\right)\left(\sum_{i=1}^{n} \sum_{j=1}^{n} w_{ij}\right)}$$

The top line with the double summation can be interpreted as follows. For every individual grid square (or zone) in turn in the area of analysis, of which there are *n*:

1. Focus on a local area that you have previously specified (for example, a 3 × 3 or 5 × 5 window).

2. Compute the value of the grid square minus the global data mean $(y_i - \bar{y})$.

3. Now look around your local neighbourhood. In a rook's case analysis, the weight w_{ij} of a touching grid square to y_i is 1, but otherwise 0; alternatively, samples within a certain distance band are weighted whilst those outside the band are weighted as zero.

4. For every individual grid square (or zone) in the specified local neighbourhood in turn in the area of analysis (of which there are *n*), compute the difference between a contributing cell and the global mean $(y_j - \bar{y})$.

5. For every individual grid square (or zone) in the specified local neighbourhood, multiply the weight for that square, the result of step 2 above and the result of step 4.

6. Total the individual results for the neighbourhood zone from step 5 $\left(\text{or } \sum_{j=1}^{n} w_{ij} z_i z_j\right)$.

7. Add the result from step 6 to your running total for all grid squares $\left(\text{or } \sum_{j=1}^{n} w_{ij} z_i z_j\right)$.

The bottom line of the formula uses a similar summation system; you can recycle the values for 2 for each grid square *i* above as you go along, by squaring the value and adding it to an overall running total [8]. Similarly, you can keep a running overall total of the weights you use for each grid square in step 5 [9].

geographical literature include **Geary's C** (also known as Geary's contiguity ratio, Geary's ratio, or the Geary index (Geary 1954)) and the **joint counts method**.

Geary's *C* ranges between 0 and 2, where 1 indicates no spatial autocorrelation. This is less easy to interpret than Moran's *I* measure in that values less than 1 indicate positive spatial autocorrelation and values greater than 1 indicate increasing levels of negative spatial autocorrelation. The joint counts method is more widely encountered in ecological rather than geographical science, but is only suitable for presence/absence data across a regular grid; in this sense the methods we have focused on here in more detail are more sophisticated and better suited for a range of tasks and data types. If you would like to explore the joint counts method in more detail, a useful worked example can be found in de Smith *et al.* (2007, p.215).

● Measuring local association

Recall that a global measures of spatial autocorrelation is one that gives a single summary measure of the patterns of association for the whole study region. The problem with this approach is it can conceal more localised patterns within the region. To detect these, local measures are used. Measuring local association in a data set is a useful process, both as a test that the standard statistical assumption of (spatial) independence have been met and as an exploratory process in its own right. A comparison of results using a variety of local and global statistics can for example be used to reveal MAUP issues within an analysis and help you determine that elusive factor, the 'most appropriate scale', for your analysis. Often, local statistics are computed as an accompaniment to the semi-variogram. We are also increasingly seeing spatial structure explicitly built into more complex statistical methods in the advanced literature.

Over the past 15 years there has been an increased trend for research articles, across both human and physical aspects of geography and environmental science, to use the Moran's I_i and Getis's G_i methods we highlight in this section. Once you have looked into the methods in more detail within the following sections, why not follow up a few examples in an area of interest to you to see the variety of ways in which they can be applied? (See Table 8.3 below.)

● Moran's I_i

The local variant of Moran's I, known as I_p is outlined in Anselin (1995). The local values of I sum to the global Moran's I within the same data set; as such, equation 8.11 below is a simplification or sub-component of equation 8.7. Computation of the value uses a rook's case neighbourhood (Section 8.1) as follows:

$$I_i = z_i \sum_{j=1}^{n} w_{ij}\, z_j, \quad j \neq i \tag{8.11}$$

where z_j are differences in the value of the variable y from its global mean; for the point j, for example, $(y_j - \bar{y})$.

A modified version of this measure can also be found within the wider literature, in which I_i is divided by the sample variance s^2.

Figure 8.10 demonstrates how to compute I_i for a small grid; this is a visual sub-component of the formula already unpacked for the global form of the statistic, and it may help to look at both explanations together if you struggled to understand how the global form worked.

When I_i is high and positive, it means that the location under investigation has similarly high (or low) values as its neighbours, so forming a cluster. Spatial outliers (a high negative value) occur when a high value rises in a generally low-value neighbourhood, or vice versa. The four general outcomes (high–high, low–low, high–low, low–high) are sometimes known as the **cluster or outlier type** (Figure 8.11). The total of all local I_i values across a grid are bounded by −1 and +1; individually, however, they are not when computed using equation 8.11.

As Anselin (1995) notes, it is possible to construct a standardised form of the local Moran's I so that, assuming the data conform to the normal distribution, its

22	21	23
25	22	23
21	25	22

Global data mean = 22.667
Global data variance = 2.25

Weighting values: 0.25 for each pale blue square (weights must total 1 over 4 contributing squares in the rook's case for a 3×3 window)

(a)

−0.667 (0)	−1.667 (0.25)	0.333 (0)
2.333 (0.25)	−0.667	0.333 (0.25)
−1.667 (0)	2.333 (0.25)	−0.667 (0)

Sum of weights = (−1.667×0.25) + (2.333×0.25) + (0.333×0.25) + (2.333×0.25) = 0.833

Original I_i = −0.667×0.833 = −0.556

Modified I_i = −0.556/2.25 = −0.25

(b)

Figure 8.10 Computation of Moran's I^* for the central grid square, using a rook's case neighbourhood and exemplifying the case where the grid is the entire data set in question: (a) original grid values (y_j); (b) difference between original grid values and the global mean $(y_j - \overline{y})$, or z_j, with standardised weight w_{ij} in brackets. In this case, I_i is −0.556, a negative value indicating that neighbourhood values tend to be dissimilar

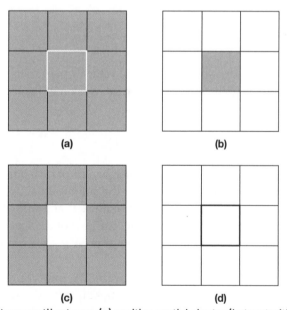

(a)   **(b)**

(c)   **(d)**

Figure 8.11 Cluster or outlier types: (a) positive spatial cluster (hotspot with high positive Z score); (b) spatial outlier with low negative Z score; (c) spatial outlier with low negative Z score; and (d) negative spatial cluster (coldspot) with high positive Z score.

significance level can be tested. In this context, the null hypothesis is that the I_i values occur randomly. Thus, Moran's I_i values may be transformed to z scores in which values greater than 1.96 or smaller than −1.96 indicate spatial autocorrelation that is significant at the 5% level. However, if the raw data are heavily skewed then the probability distribution of the local Moran's I may well not be normal; to cope with this situation, an alternative method known as conditional permutation is often preferred. If you see reference to a p value, as well as a z value, in the software you may be using to compute I_i this is the pseudo significance value computed using the conditional permutation method (See also Chapter 6, section 6.8).

Moran's I_i has been used for a wide range of applications, both practical and theoretical in nature. As Anselin (1995) points out, the direct linkage between local indicators to the global measure of spatial autocorrelation means that 'the decomposition of the latter into its observation-specific components becomes straightforward, thus enabling the assessment of influential observations and outliers' (p. 112). It is this characteristic that allowed Nakhapakorn and Jirakajohnkool (2006) for example to explore local measures of spatial autocorrelation to seek outliers in the context of an analysis to explore the spatial factors affecting a dengue fever epidemic in the Sukhothai province of Thailand.

Moran's I_i can also potentially indicate local spatial clusters in cases where no overall global spatial autocorrelation is present in a data set (Anselin 1995).

Getis's G_i

The formula for computing the local form of the Getis statistic is

$$G_i = n \frac{\sum_{j=1}^{n} w_{ij} z_j}{\sum_{j=1}^{n} z_j}, \quad i \neq j \tag{8.12}$$

where z_j are differences in the value of the variable y from its global mean. You may also see in the literature a further version of the local G statistic, referred to as G_i^*:

$$G_i^* = n \frac{\sum_{j=1}^{n} w_{ij} z_j}{\sum_{j=1}^{n} z_j} \tag{8.13}$$

In the first version, only the neighbours of each point are considered in the calculation, while the second also includes the point itself. G_i^* is currently the more commonly used form of the G_i index. As with Moran's I_p and with similar statistical caveats, the z score of the index indicates whether the level of the score is significantly different from that expected of a random distribution. A z score close to zero indicates no significant cluster is present.

Comparing equation 8.10 with 8.12 and 8.13, you can see that, unlike Moran's I_p the individual components are not directly related to the global G statistic. This makes Moran's I_i the preferred approach for sensitivity analyses of potential outliers.

Getis's G_i statistic is less widely used than Moran's I_p but retains a following due to its ability to retain a sense of the values in the data set in addition to the level of spatial autocorrelation. As with Moran's I_p it is able to identify **hotspots,** local areas of non-stationarity within the data. In their original exposition of the G_i statistic, for example, Getis and Ord (1992) demonstrated how local pockets of spatial dependence were revealed in county data for sudden infant death syndrome by county in North Carolina when using the local form of G. This paper is also a reminder that spatial association in a data set need not imply causality. Sudden infant death syndrome remains not fully understood, but may to some degree relate to socio-economic factors such as smoking rather than spatial connectivity per se. However, those factors themselves vary geographically so an understanding of geography remains important. A wider range of examples of both Moran's I_i and Getis's G_i, in both social and environmental contexts, is highlighted within Table 8.3. Importantly, a number of papers compute both Moran's I and Getis's G, as the two measures offer different aspects to an analysis.

Key concept 8.8 Hotspot

A **hotspot** is a cluster of high or low values in a data set. It is often used to infer positive spatial autocorrelation in particular; in other words, a cluster with high values in comparison to its surrounding area. See Figure 8.11.

For example, it could be an area with high crime rates, mortality or disease (Tiwari et al. 2006).

Table 8.3 **Exemplar uses of local measures of spatial association in geography and environmental science**

Subject	Authors	Statistic
Spatial radiometric stability of test sites used for calibrating Earth observation sensors	Bannari et al. (2005)	G_i^*
Probable local concentration of slope failure in Northern Taiwan	Chu et al. (2009)	G_i^*
Associations between snow cover and atmospheric circulation	Derkson et al. (1998)	G_i^*
Textural aid in multispectral image classification	Emerson et al. (2005)	I_i
Local pockets of spatial dependence for sudden infant death syndrome	Getis and Ord (1992)	G_i^*
Clustering in breast, lung and colorectal cancer	Jacquez and Greiling (2003)	I_i
Spatial clustering of forest cover in remotely sensed data	Wulder and Boots (1998)	G_i^*
Hotspots of lead pollution in urban soils	Zhang et al. (2008)	I_i
Spatial pattern of mosquito vectors of the Ross River and Barmah Forest virus	Jeffery et al. (2002)	G_i^*
Hotpots in crime in Nottinghamshire	Ratcliffe and McCullagh (2001)	G_i^*
Spatial clustering of employment	Ceccato and Persson (2002)	G_i^*
Accessibility of public playgrounds	Talen and Anselin (1998)	I_i

8.6 Conclusion

In the following chapter we develop your knowledge of Moran's *I* further. In particular we look more closely at the nature of spatial weights, and definitions of neighbourhoods. We also outline how we can incorporate notions of global and local autocorrelation within statistical models to provide tools explicitly designed for spatial analysis.

Key points

Spatial dependence, or spatial autocorrelation, is a key issue in quantitative geographical enquiry for a variety of reasons:

- Failure to assess the effect of spatial autocorrelation in the context of a statistical study means that you may be violating underlying assumptions of independence in your data that are key to your method.
- Spatial autocorrelation, through the MAUP, means that the results of an analysis may be entirely zone or scale specific.
- Failure to consider the ecological fallacy leads both to varied results by scale and to inappropriate inference regarding individual behaviour; consider using a mix of qualitative and quantitative methods if you are concerned about the latter in your own research.
- Assessing global measures of spatial autocorrelation is a step towards reviewing whether a standard statistical analysis is appropriate or whether an adjusted method (or sampling regime) is required.
- Methods to assess spatial autocorrelation, such as the variogram or Moran's I_i or Getis G_i^* can be very useful ways of getting to know your subject better in addition to providing help in evaluating an appropriate scale for analysis and mitigating statistical problems.

References

Anselin, L. (1995) Local indicators of spatial association – LISA. *Geographical Analysis,* 27, 93–115.

Bannari, A., Omari, K., Teillet, P.M. and Fedosejevs, G. (2005) Potential of Getis statistics to characterize the radiometric uniformity and stability of test sites used for the calibration of Earth observation sensors. *IEEE Transactions on Geoscience and Remote Sensing,* 43, 2918–2926.

Brunsdon, C., Fotheringham, A.S. and Charlton, M.E. (1996) Geographically weighted regression: a method for exploring spatial nonstationarity. *Geographical Analysis,* 28, 281–298.

Burrough, P.A. and McDonnell, R.A. (1998) *Principles of Geographical Information Systems,* Oxford: Oxford University Press.

Ceccato, V. and Persson, L.O. (2002) Dynamics of rural areas: an assessment of clusters of employment in Sweden. *Journal of Rural Studies,* 18, 49–63.

Chu, C.-M., Tsai, B.-W. and Chang, K.-T. (2009) Integrating decision tree and spatial cluster analysis for landslide susceptibility zonation. *World Academy of Science, Engineering and Technology,* 59, 479–483.

Cliff, A.D. and Ord, J.K. (1969) The problem of spatial autocorrelation. In A.J. Scott (eds), *London Papers in Regional Science 1, Studies in Regional Science,* London: Pion, pp.25–55.

Clifford, P., Richardson, S. and Hémon, D. (1989) Assessing the significance of the correlation between two spatial processes. *Biometrics,* 45, 123–134.

Connolly, P. (2006) Summary statistics, educational achievement gaps and the ecological fallacy. *Oxford Review of Education,* 32, 235–252.

Couclelis, H. and Getis, A. (2000) Conceptualising and measuring accessibility within physical and virtual spaces. In D.G. Janelle and D.C. Hodge (eds), *Information, Place and Cyberspace: Issues in Accessibility,* Berlin: Springer, pp.15–20.

Dark, S.J. (2004) The biogeography of invasive alien plants in California: an application of GIS and spatial regression analysis. *Diversity & Distributions,* 10, 1–9.

Dark, S.J. and Bram, D. (2007) The modifiable areal unit problem (MAUP) in physical geography. *Progress in Physical Geography,* 31, 471–479.

De Smith, M., Goodchild, M.F. and Longley, P.A. (2007) *Geospatial Analysis: A Comprehensive Guide to Principles, Techniques and Software Tools,* Leicester: Matador.

Derksen, C., Wulder, M., LeDrew, E. and Goodison, R. (1998) Associations between spatially autocorrelated patterns of SSM/I-derived prairie snow cover and atmospheric circulation. *Hydrological Preview,* 12, 2307–2316.

Diniz-Filho, J.A.F., Bini, L.M. and Hawkins, B. (2003) Spatial autocorrelation and red herrings in geographical ecology. *Global Ecology and Biogeography,* 12, 53–64.

Emerson, C.W., Lam, N.S.-N. and Quattrochi, D.A. (2005) A comparison of local variance, fractal dimension and Moran's I as aids to multispectral image classification. *International Journal of Remote Sensing,* 26, 1575–1588.

Fotheringham, A.S. and Wong, D.W.S. (1991) The modifiable areal unit problem in multivariate statistical analysis. *Environment and Planning A,* 23, 1025–1044.

Geary, R.C. (1954) The contiguity ratio and statistical mapping. *Incorporated Statistician,* 5, 115–145.

Getis, A. and Ord, K. (1992) The analysis of spatial association by use of distance statistics. *Geographical Analysis,* 24, 189–206.

Gudgin, G. and Taylor, P.J. (1979) *Seats, Votes and the Spatial Organization of Elections,* London: Pion.

Hay, G.J., Marceau, D.J., Dube, P. and Bouchard, A. (2001) A multiscale framework for landscape analysis: object-specific analysis and upscaling. *Landscape Ecology,* 16, 471–490.

He, Z., Zhao, W. and Chang, X. (2007) The modifiable areal unit problem of spatial heterogeneity of plant community in the transitional zone between oasis and desert using semivariance analysis. *Landscape Ecology,* 22, 95–104.

Heywood, I., Cornelius, S. and Carver, S. (2006) *An Introduction to Geographical Information System,* 3rd edn, Harlow: Pearson Education.

Jarvis, C.H. (2000) Insect phenology: a geographical perspective. PhD thesis, University of Edinburgh.

Jarvis, C.H. and Baker, R.H.A. (2001) Risk assessment for non-indigenous pests: 1. Mapping the outputs of phenology models to assess the likelihood of establishment. *Diversity & Distributions,* 7, 223–235.

Jacquez, G.M. and Greiling, D.A. (2003) Local clustering in breast, lung and colorectal cancer in Long Island, New York. *International Journal of Health Geographics,* 2. Available at: http://www.ij-healthgeographics.com/content/2/1/3, accessed 10 April 2010.

Jeffery, J.A., Ryan, P.A., Lyons, S.A., Thomas, P.T. and Kay, B.H. (2002) Spatial distribution of vectors of Ross River Virus and Barmah Forest virus on Russell Island, Moreton Bay, Queensland. *Australian Journal of Entomology,* 41, 329–338.

Lennon, J.J. (2000) Red-shifts and red herrings in geographical ecology. *Ecography,* 23, 101–113.

Longley, P.A., Goodchild, M.F., Maguire, D.J. and Rhind, D.W. (2011) *Geographical Information Systems and Science,* 3rd edn, Hoboken, NJ: Wiley.

Miller, H. (2004) Tobler's First Law and spatial analysis. *Annals of the Association of American Geographers,* 94, 284–289.

Moran, P.A.P. (1950) Notes on continuous stochastic phenomena. *Biometrika,* 37, 17–33.

Nakhapakorn, K. and Jirakajohnkool, S. (2006) Temporal and spatial autocorrelation statistics of Dengue Fever. *Dengue Bulletin,* 30, 177–183.

Openshaw, S. (1984) Ecological fallacies and the analysis of areal census data. *Environment and Planning A,* 16, 17–31.

Ord, J.K. and Getis, A. (1995) Local spatial autocorrelation statistics: distributional issues and an application. *Geographical Analysis,* 27, 286–306.

Ratcliffe, J.H., and McCullagh, M.J. (2001) Chasing ghosts? Police perception of high crime areas. *British Journal of Criminology,* 41, 330–341.

Robinson, W.S. (1950) Ecological correlations and the behavior of individuals. *American Sociological Review,* 15, 351–357.

Rossiter, D.J., Johnston, R.J. and Pattie, C.J. (1999) *Redrawing of the UK's Map of Parliamentary Constituencies,* Manchester: Manchester University Press.

Talen, E. and Anselin, L. (1998) Assessing spatial equality: an evaluation of measures of accessibility to public playgrounds. *Environment and Planning A,* 30, 595–613.

Tiwari, N., Ahikari, C.M.S., Tewari, A. and Kandpal, V. (2006) Investigation of geo-spatial hotspots for the occurrence of tuberculosis in Almora district, India, using GIS and spatial scan statistic. *International Journal of Health Geographics,* 5. Available at: http://www.ij-healthgeographics.com/content/5/1/33, accessed 10 April 2010.

Webster, R. and Oliver, M.A. (2007) *Geostatistics for Environmental Scientists,* 2nd edn, Chichester: Wiley.

Weissmann, G.S. and Fogg, G.E. (1999) Multi-scale alluvial fan heterogeneity modelled with transition probability geostatistics in a sequence stratigraphic framework. *Journal of Hydrology,* 226, 48–65.

Wulder, M. and Boots, B. (1998) Local spatial autocorrelation characteristics of remotely sensed imagery assessed with the Getis statistic. *International Journal of Remote Sensing,* 19, 2223–2231.

Zhang, C., Luo, L., Xu, W. and Ledwith, V. (2008) Use of local Moran's I and GIS to identify pollution hotspots of Pb in urban soils of Galway, Ireland. *Science of the Total Environment,* 398, 212–221.

Exploring spatial relationships

Chapter overview

Multiple regression is not a very geographical method of analysis. This is most obvious when interesting patterns are found in the residuals but these same patterns are treated as a nuisance, a violation of the assumption the residuals are independent.

This is not very satisfactory from a geographical perspective. It may well be that the processes creating the patterns have a geographical cause. For example, they may be formed by the interaction between people and places or between species and their environment. Alternatively, what happens at one location could directly be affected by what happens around it. Context and surroundings matter.

This chapter is about treating *where* something happens as useful information that may help explain *what* is happening. The central idea is when we find geographical patterns in data and there is evidence to suggest that they did not arise by chance, so it would be better to explore and model the cause of the patterns than to treat them as an inconvenience.

Methods to do so include spatial regression models, geographically weighted regression (GWR) and multilevel modelling.

9.1 Introduction

In the previous chapter we looked at ways to detect geographical patterns in data. Such patterns violate assumptions of independence underpinning many statistical tests.

The related problem is that many of the statistical methods used to analyse geographical data are not, themselves, geographical. This takes us back to the definition of geographical data given in Chapter 1, Section 1.5. Geographical data are records of what has happened at some location on or near the Earth's surface. Whilst the key characteristic of a geographical data set is that it records what happens *and where,* the locational information is of no direct relevance to the statistical test (though it may become relevant afterwards for the interpretation of the result and to check that assumptions of independence have been met).

To clarify this, look at Figures 9.1(a) and 9.1(b). Both show an index of non-participation in UK higher education institutions calculated for census tracts in London for the period 2002–2005. The higher the index value, the less probably a young adult aged 21 years or under and living within the area will go to a university or college.

Figure 9.1(a) shows the true data for London. If you were looking at this map in a geographical information system (GIS) it would be linked to an attribute table where the rows in the table give the non-participation index together with other data for each

(a) Actual pattern of non-participation

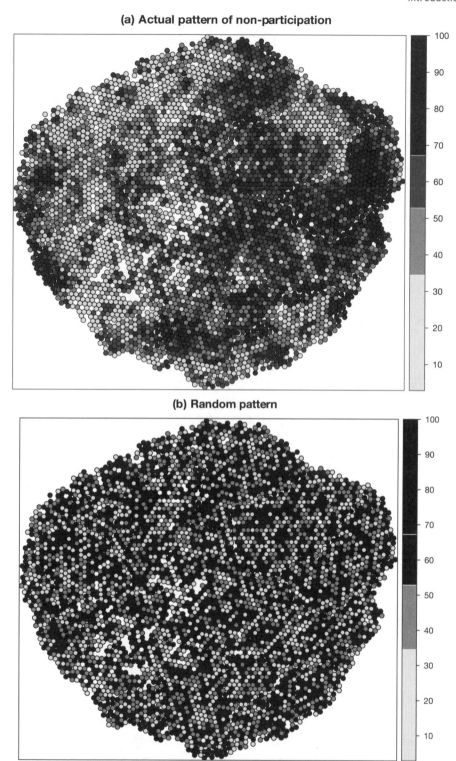

(b) Random pattern

Figure 9.1 Mapping an index of non-participation in UK higher education for London neighbourhoods. (a) This shows the actual patterns of non-participation. (b) Here the pattern is random and arises from randomly moving the data around

Data source: Based on Harris *et al.* (2010).

Table 9.1 Regression analysis of the attribute data behind Figures 9.1(a) and 9.1(b). The results are the same. The variables have been standardised (see Chapter 7, Section 7.12)

	(a)				(b)			
	Estimate	se	t	p	Estimate	se	t	p
(Intercept)	0.000	0.009	0.000	1.000	0.000	0.009	0.000	1.000
% adults in the area with no educational qualifications (standardised)	0.640	0.010	63.7	0.000	0.640	0.010	63.7	0.000
% households with four or more cars (standardised)	−0.276	0.010	−26.4	0.000	−0.276	0.010	−26.4	0.000
% population Indian, Pakistani or Bangladeshi (standardised)	−0.263	0.009	−28.2	0.000	−0.263	0.009	−28.2	0.000
Average attainment score in final year of compulsory education (standardised)	−0.137	0.011	−12.3	0.000	−0.137	0.011	−12.3	0.000
R^2	0.614				0.614			
$F_{(4,4692)}$	1863			0.000	1863			0.000
AIC	8873				8873			

of the census tracts on-screen. Figure 9.1(b) is what happens if the rows of the attribute table get accidently rearranged, leaving each record associated with the wrong place on the map. Hence, whereas Figure 9.1(a) reflects the actual social geography of London, Figure 9.1(b) shows a random patterning.

Next we compare the results when a regression model is fitted first to the original attribute table and then to its reordered counterpart. Notably, the results are the same (Table 9.1). As far as the analysis is concerned the data are identical. The order of the rows is immaterial and where the data are located is irrelevant because location is not an input into the regression.

Yet, clearly the maps are not the same: the patterns in the data differ. Rather than discarding the geographical information, we want to use it. Knowing where something happens and what happens around it is useful for understanding why the patterns have formed. It sheds light on the underlying processes. This chapter is about methods of regression-based analysis that incorporate and make use of that geographical knowledge.

 ## 9.2 Mapping with cartograms

Those who are familiar with the geography of London may be suspicious of our maps in Figure 9.1. Further doubt arises from knowing that census tracts are not round. Before moving to more geographical methods of analysis, some explanation is required about why we are presenting the data in this way.

In most maps the size of each region, neighbourhood or environmental zone is scaled in proportion to its area. Larger neighbourhoods occupy more of the map space than smaller ones. This makes a lot of sense, it tells you which places cover which terrain and it makes the map useful for navigation and surveying. However, neighbourhoods rarely have natural boundaries. Census tracts are designed for administrative purposes. The reason why some are smaller is because they have a higher

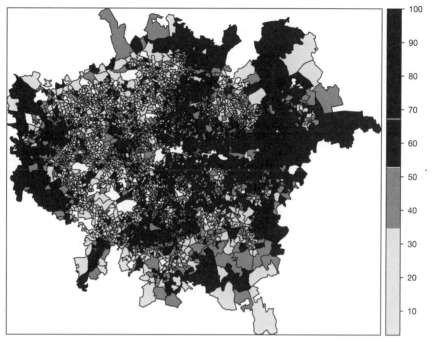

Figure 9.2 The index of non-participation values shown on a conventional choropleth map. It is now difficult to see the values for the central regions

concentration of people living there. The population density is higher. Often we are not interested in the places, per se, but in the people living within them. It is not ideal for conventional maps to give least space to the places where, from a social point of view, most is happening.

Figure 9.2 shows the true participation index for census tracts in London as a conventional choropleth map. The map becomes dominated by the peripheral areas with more green space, whereas the central regions are hard to see. Whilst the geography of this map remains important – it is the one we use to determine which places border others in the analyses that follow – for conveying the patterns of non-participation, a map that gives more space to the higher density areas is better. It is a more effective visualisation tool.

A cartogram is a map where the size of the areas on it are scaled in proportion to something other than land area – by population density, for example. This is effective in political geography where in broadly two-party systems it helps to avoid the impression of one party winning overwhelmingly when, in fact, its vote was concentrated in larger but sparsely populated rural areas (see http://www.geog.ucsb.edu/~sara/html/mapping/election/map.html on the 2000 US presidential elections). Cartograms can also be used to give focus to places of most concern to a particular study – for example, those with the highest infant mortality rates, lowest life expectancy or greatest greenhouse gas emissions (Monmonier 2010; Thomas and Dorling 2007; www-personal.umich.edu/~mejn/cartograms/).

For our maps we have produced a very simple cartogram based on an algorithm by Dorling (1996) but choosing to give each census tract equal representation on the map. There are a number of free tools available to produce cartograms, including ScapeToad (http://scapetoad.choros.ch/) and the Cartogram creator for Quantum GIS (www.qgis.org and www.ftools.ca).

> **Key concept 9.1 Cartograms**
>
> A **cartogram** is a map in which the areas are scaled in proportion to some variable other than geographical distance. In this way the areas with the greatest amount of whatever the study is interested in (people, health, environmental pollution, wealth, etc.) become most prominent on the map and are the ones to which the eye is drawn.

9.3 Spatial analysis and spatial autocorrelation

Spatial analysis is a set of methods whose results are not invariant to where the data are collected (Longley *et al.* 2010). In other words, change the information about where the data are collected and the results will change too. Under this definition, regression and many other methods of statistical analysis are identified as non-spatial.

This is neither a criticism of conventional regression analysis nor a suggestion it should not be used for geographical research. It is more a comment about using the right tool for the job.

Looking at Figure 9.1(a), there is evidence of a geographical pattern of non-participation in higher education within London. Whilst this does not necessarily mean the assumption of residual independence has been violated, it would not be surprising if the residuals did reflect the pattern. Certainly, it will need checking.

To do so, knowing where the data were collected becomes important again. The simplest check is what we did in Chapter 7: plot the residuals on a map and look at it. A better approach, less susceptible to subjective judgement, is to compare the residual value for each place with those of its contiguous neighbours. This was the thinking behind the Moran plot shown in Chapter 7, Figure 7.11. Unfortunately the plot becomes cluttered when, in the case of the non-participation data, we have $n = 4697$ observations to draw on it.

Another option is to summarise what the plot shows in terms of a correlation statistic. This will compare the similarity of each place with its neighbours, averaging across the study region. A way to do so was introduced in Chapter 8 and is Moran's I statistic (Moran 1948; 1950). This gives a negative value when neighbouring places have, on average, contrasting values, and a positive value when neighbours have, on average, similar values. A value of zero suggests there is no marked spatial pattern.

What we mean by contrasting can be represented in terms of a checkerboard effect. This is shown in Figure 9.3(a) and is an example of negative spatial autocorrelation where proximate values differ. This can arise owing to processes of social, economic or ethnic separation/segregation. A more common occurrence is positive spatial occurrence, shown in Figure 9.3(b), where alike values cluster together. Recall that it is positive spatial autocorrelation that is described by Tobler's 'first law of geography' (Chapter 1, Key concept 1.5). Figure 9.3(c) shows an entirely random pattern.

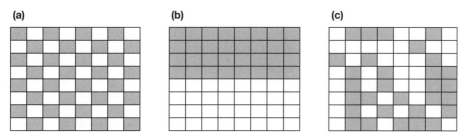

Figure 9.3 Examples of (a) negative spatial autocorrelation; (b) positive spatial autocorrelation; and (c) no spatial autocorrelation (a random pattern)

The regression residuals for the model of non-participation in higher education give a Moran value of $I = 0.157$. Using an exact test (Hepple 1998; Tiefelsdorf 1998) we find it to be significant at a 99.9% confidence level, providing evidence of positive spatial autocorrelation between the residuals. This follows from the pattern of non-participation that we saw in the map and the failure of the model fully to explain that pattern. (The Moran value for the randomly relocated data is $I = 0.009$ and is not significant at even a 90% confidence level.)

The equation for calculating the Moran value is given by

$$I = n \frac{\sum_{i=1}^{n} \sum_{j=1}^{n} w_{ij}(y_i - \overline{y})(y_j - \overline{y})}{\left(\sum_{i=1}^{n} (y_i - \overline{y})^2\right)\left(\sum_{i=1}^{n} \sum_{j=1}^{n} w_{ij}\right)} \tag{8.7}$$

Be aware of the risk of confusion. Here y means simply 'a value', not necessarily the Y variable in a regression relationship. The values we used were the regression residuals.

Recall also that, although the equation appears formidable, all it does is look at how the value for one place (denoted by the notation y_i) correlates with a value for another place (y_j), working through all places in sequence ($\sum_{i=1}^{n} \sum_{j=1}^{n}$), comparing all places

---

Key concept 9.2 Spatial analysis

Spatial analysis is a set of methods whose results are not invariant to where the data are collected. That means they are sensitive to geographical patterns within the data – if the pattern changes so too does the result. The methods can be used to search for and summarise the patterns, and to measure processes that create the geographical patterning.

Key concept 9.3 Moran's *I*

Moran's *I* is a measure of how correlated neighbouring values are across a study region where neighbours are those that share a border, are within a certain distance of each other, where there is a flow between them, or where they are linked in some other way.

A positive value of *I* indicates positive spatial autocorrelation: neighbouring values tend to be alike. A negative value of *I* indicates dissimilarity between neighbours.

with all others. In many cases the comparison is irrelevant because the places are not neighbours and can therefore be discarded. This is achieved by setting the weight w_{ij} to zero when two places are not neighbours, and to some value greater than zero if they are (See Chapter 8 section 8.3).

9.4 Who's my neighbour? Defining a weights matrix

The calculation of Moran's I requires us to define which places are neighbours and which are not. There are many ways to define neighbours. In our example it is two places that share a common border. Look, however, at areas 1 and 5 in Figure 9.4. These meet at a corner and do not strictly share a border. Are these neighbours? Probably we would want to say yes, in which case neighbours could be defined as two places that either share a border or are touching (Queen's case contiguity, see Chapter 8). We may also wish to include areas that are not touching but are a small distance apart, allowing for digitisation errors where areas do not meet as exactly. Alternatively, we could define neighbours as all places that are within a certain distance of the centre or boundary of another (regardless of whether they share a boundary or not), or look to find the first, second, third, fourth, . . . , or whatever nearest places to another and describe those as its neighbours.

Having decided on the definition, a record of the neighbours is required. A simple way to do this is in an n by n weights matrix, where n is the number of observations (places) within the data set. Figure 9.5 shows a weights matrix for just the areas numbered in Figure 9.4. Area 1 has areas 2, 5, 9 and 10 as its neighbours, so the values at (row 1, column 2), (row 1, column 5), (row 1, column 9) and (row 1, column 10) all

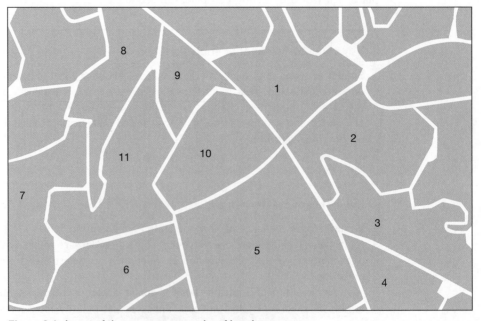

Figure 9.4 **A part of the census geography of London**

	1	2	3	4	5	6	7	8	9	10	11
1	0	1	0	0	1	0	0	0	1	1	0
2	1	0	1	0	1	0	0	0	0	1	0
3	0	1	0	1	1	0	0	0	0	0	0
4	0	0	1	0	1	0	0	0	0	0	0
5	1	1	1	1	0	1	0	0	0	1	1
6	0	0	0	0	1	0	1	0	0	1	1
7	0	0	0	0	0	1	0	1	0	0	1
8	0	0	0	0	0	0	1	0	1	0	1
9	1	0	0	0	0	0	0	1	0	1	1
10	1	1	0	0	1	0	0	0	1	0	1
11	0	0	0	0	1	1	1	1	1	1	0

Figure 9.5 A weights matrix for the areas shown in Figure 9.4

	1	2	3	4	5	6	7	8	9	10	11
1	0.00	0.25	0.00	0.00	0.25	0.00	0.00	0.00	0.25	0.25	0.00
2	0.25	0.00	0.25	0.00	0.25	0.00	0.00	0.00	0.00	0.25	0.00
3	0.00	0.33	0.00	0.33	0.33	0.00	0.00	0.00	0.00	0.00	0.00
4	0.00	0.00	0.50	0.00	0.50	0.00	0.00	0.00	0.00	0.00	0.00
5	0.14	0.14	0.14	0.14	0.00	0.14	0.00	0.00	0.00	0.14	0.14
6	0.00	0.00	0.00	0.00	0.25	0.00	0.25	0.00	0.00	0.25	0.25
7	0.00	0.00	0.00	0.00	0.00	0.33	0.00	0.33	0.00	0.00	0.33
8	0.00	0.00	0.00	0.00	0.00	0.00	0.33	0.00	0.33	0.00	0.33
9	0.25	0.00	0.00	0.00	0.00	0.00	0.00	0.25	0.00	0.25	0.25
10	0.20	0.20	0.00	0.00	0.20	0.00	0.00	0.00	0.20	0.00	0.20
11	0.00	0.00	0.00	0.00	0.17	0.17	0.17	0.17	0.17	0.17	0.00

Figure 9.6 The matrix has been row standardised (the values in each row sum to one)

equal one. The other values in the row are zero – they represent places that are not neighbours of area 1. In the same way, area 2 has areas 1, 3, 5 and 10 as its neighbours, so the values at (row 2, column 1), (row 2, column 3), (row 2, column 5) and (row 2, column 10) are all one. And so forth.

Note that it is conventional not to define an area as its own neighbour. Therefore, (row 1, column 1), (row 2, column 2), (row 3, column 3), etc., all contain zero values. Note also that in this case the matrix is symmetrical: the value at (row 1, column 2) is the same as at (row 2, column 1); the value at (row 2, column 3) is the same as at (row 3, column 2); and so forth. This is because we have defined neighbours in such a way that if one place is a neighbour of a second then the second must also be a neighbour of the first – they share a common boundary.

The matrix need not be symmetrical. One place can be one of ten closest places to a second without the reverse being true. Or, imagine a situation where people move from one place to another but the return direction is more difficult. In this case it may make little sense to treat the two places as equal neighbours.

Furthermore, there is no requirement to limit the matrix to binary (1/0; yes/no) values. We could enter the length of the boundary or some measure of, say, migration, trade or visitors between the places, as appropriate to the study.

Having defined the matrix, it is common to scale the weights in each row so they sum to one. This is described as row standardisation. The weights matrix is standardised in Figure 9.6.

Unfortunately, as the number of observations increases, the weights matrix quickly grows, requiring large amounts of memory or file space to store it. Most of it will also contain zeros since most places have few neighbours. A more parsimonious way of storing the data is in a format that can be saved and edited as a simple text (.txt) file:

11

1 4

    2   5   9   10

2 4

    1   3   5   10

3 3

    2   4   5

(etc.)

The file begins by stating there are 11 observations. In the next line, observation 1 is identified as having four neighbours: 2, 5, 9 and 10. Observation 2 also has four neighbours: 1, 3, 5 and 10. Observation 3 has three neighbours, and so forth.

Software such as GeoDa and the sp library for R (Bivand *et al.* 2008) can create weights matrices automatically from standard GIS files (namely, shapefiles, .shp). Thinking carefully about how neighbours are defined is important. Clearly it will affect the definition of the weights matrix and therefore the result of the analysis that utilises it. An element of 'trial and error' may be required, noting how sensitive the results are to changes.

> **Key concept 9.4 The weights matrix**
>
> The (spatial) **weights matrix** is a way of encoding the relationship between areas shown on a map. For example, whether they share a border, how long that border is or the flow of goods across it. The specification of the weights matrix affects directly the results of statistics such as Moran's *I* that use it.

9.5 Spatial error model

Imagine that before building the regression model summarised by Table 9.1(a) we first used a Moran test to quantify the geographical patterning evident in the dependent variable, the index of non-participation in higher education. Doing so gives $I = 0.501$, significant at greater than a 99.9% confidence level. Knowing this, we might have anticipated spatial correlations in the regression residuals and sought a method to accommodate them.

The spatial error model provides a way. It takes the form

$$y = \beta_0 + \beta_1 x_1 + \beta_2 x_2 + \cdots + \lambda \mathbf{W} \xi + \varepsilon \tag{9.1}$$

Again, do not be put off by the notation (which follows Ward and Gleditsch 2008). It is the same as a conventional multivariate regression model except the residual errors have been split into two parts: those that are spatially correlated, ξ; and those that are not, ε. The **W** is the weights matrix discussed in the preceding section. When we look to see if the errors are spatially correlated we are asking whether neighbouring places have similar residual errors. The weights matrix defines the neighbours. The task is to estimate the parameter λ (lambda), which is a measure of spatial autocorrelation. If $\lambda \neq 0$ there may be a pattern of spatial dependence between the errors. The null hypothesis of it arising by chance must then be tested.

The results of fitting the spatial error model to the non-participation data are shown in Table 9.2, first for the correctly located data and then for the randomly relocated data. The results for the former reveal significant positive spatial autocorrelation between the errors, $\lambda = 0.271$, significant at greater than a 99.9% confidence level. This means that we have, in effect, less than the $n = 4697$ independent observations we thought we had and this reduction increases our uncertainty in the data. Because of this, the standard error associated with each estimated effect has risen (compare the standard errors shown under (a) in Table 9.2 with those in Table 9.1).

Overall, the spatial error model fits the model better than the standard regression model with its (here, erroneous) assumption of independent and identically distributed residuals. The improvement is reflected in the decreased Akaike information criterion (AIC) score, down to 8670 from 8873 previously (the lower the AIC score, the better) AIC is a goodness-of-fit measure (see Chapter 7, section 7.12).

By contrast, the randomly relocated data contains no significant spatially correlated errors ($\lambda = 0.018$; $p = 0.383$) and neither the estimates nor their standard errors change noticeably. In fact, the AIC score for the spatial error model is marginally worse (greater) than for the conventional regression model. In this case, the lack of spatial autocorrelation does not warrant the additional complexity of the spatial error model.

What is important is that the results in Table 9.2 do now differ. The two data sets have different locational information. Since the spatial error model is an example of spatial regression modelling, changing the locational information changes the results.

Table 9.2 Spatial error models of the attribute data behind Figures 9.1(a) and 9.1(b). Unlike for the standard regression model, the results now differ. The correctly located data give significant positive spatial autocorrelation in the errors ($\lambda = 0.271$; $p = 0.000$). The variables are standardised

	(a)				(b)			
	Estimate	se	z	p	Estimate	se	z	p
(Intercept)	0.000	0.012	0.024	0.981	0.000	0.009	0.001	0.999
% adults in the area with no educational qualifications (standardised)	0.628	0.012	53.1	0.000	0.640	0.010	63.8	0.000
% households with four or more cars (standardised)	−0.273	0.012	−23.6	0.000	−0.276	0.010	−26.5	0.000
% population Indian, Pakistani or Bangladeshi (standardised)	−0.261	0.012	−22.4	0.000	−0.263	0.009	−28.2	0.000
Average attainment score in final year of compulsory education (standardised)	−0.122	0.011	−10.8	0.000	−0.137	0.011	−12.2	0.000
λ	0.271	0.019	14.3	0.000	0.018	0.021	0.873	0.383
AIC	8670				8874			

9.6 Spatial lagged *y* model

As we have seen, the spatial error model is useful for detecting and quantifying spatial correlations in the regression residuals, and for avoiding underestimation of each coefficient's standard error (an underestimation that then underestimates the *p* value, potentially leading us to conclude that a relationship is significant at a given level of confidence when it is not). However, the model provides no explanation as to why those spatial correlations exist. It says nothing about the processes that generated them.

A way forward is to treat the model as under-parameterised; to conclude that it does not yet contain all the variables required to explain the patterning of the dependent variable. The hope is that once there are enough explanatory variables, the patterns in the data will be fully explained and the spatial correlations in the residuals will disappear. It is possible but it may also be wishful thinking.

Another is to treat the geographical patterning as arising from a process that itself is geographical and that the data capture in some way. For example, we may want to argue that people interact so what happens in one place will affect what happens nearby. The same is true of the interaction between species of flora and fauna and how ecosystems evolve.

Returning to our model of (non-)participation in higher education, we may argue that whether people go to university or not is, at least in part, a measure of local attitudes to and experiences of education – that it is influenced by peers, role models and local mentors. Cultural experiences and expectations are not so neatly self-contained within the boundaries of each census tract that we could credibly believe that the people in one area live in isolation from surrounding places (especially as the boundaries of census tracts tend to be arbitrary in the UK). As such, we can hypothesise that non-participation in one area has an effect that spills over into surrounding areas.

We can test the hypothesis using a spatially lagged *y* model, also known as a spatial autoregressive model. The model has the form

$$y = \beta_0 + \beta_1 x_1 + \beta_2 x_2 + \cdots + \rho \mathbf{W}\mathbf{y} + \varepsilon \tag{9.2}$$

The key is to understand what this is saying: the value of the dependent variable is explained not just by the various predictor variables ($x_1, x_2, \ldots$) but also by the value of the dependent variable in neighbouring places ($\mathbf{W}\mathbf{y}$). The parameter, ρ (rho), measures the **overspill effect**.

The results for the spatially lagged *y* model are shown in Table 9.3. The estimates of each predictor variable's effect upon the dependent variable change slightly from the spatial error model and the AIC score suggests some improvement to the overall fit. Most notably, the estimate of ρ and its associated *p* value suggest that the rate of non-participation in one area is significantly related to the rate of non-participation in neighbouring areas. That effect is of similar magnitude to the percentage of households with four or more cars.

The spatial error model and the spatially lagged *y* model can be fitted in GeoDa and by using the spdep library for R. The models are drawn from the **spatial econometrics**

Table 9.3 The results of a spatially lagged y model predicting rates of non-participation in higher education in London. The variables are standardised.

	Estimate	se	z	p
(Intercept)	0.003	0.009	0.320	0.749
% adults in the area with no educational qualifications (standardised)	0.535	0.010	44.4	0.000
% households with four or more cars (standardised)	−0.243	0.010	−23.4	0.000
% population Indian, Pakistani or Bangladeshi (standardised)	−0.217	0.010	−22.8	0.000
Average attainment score in final year of compulsory education (standardised)	−0.122	0.011	−11.2	0.000
ρ	0.224	0.015	15.3	0.000
AIC	8637			

literature (in which other types of model can be found). As Pace and LeSage (2010, p.245) observe:

> A long-running theme in economics is how individuals or organizations following their own interest result in benefits or costs to others. These benefits or costs are labelled externalities and often termed spillovers in a spatial setting. A technological innovation provides an example of a positive externality of spillover while pollution provides an example of a negative externality of spillover [. . .] Spatial econometric models can quantify how changes in explanatory variables of the model directly impact individuals, regions, local governments, etc., as well as the associated spillover impacts. Quantifying these effects provides useful information for policy purposes.

For an introduction to spatial econometrics and spatial regression models see Ward and Gleditsch (2008). For a comprehensive treatment see LeSage and Pace (2008).

Key concept 9.5   Spatial econometrics

The **spatial error model** and the **spatially lagged y model** are examples of spatial regression models drawn from the field of **spatial econometrics**. This is an interdisciplinary field of economists, geographers and statisticians that focuses on the development of (regression) models and tests for application to geographical data and the testing of economic theory (Hepple 2001).

9.7   Geographically weighted regression (GWR)

A complementary but different approach to the spatially lagged y model is to shift attention from the dependent variable to how the effects of the independent variables vary over the study region.

Consider the situation in Figure 9.7(a) where we have attribute data for each of the points shown. The points could be the actual location where the data are collected or represent the centre of an area that the data are about – for example, the centre of a census tract.

As an example, imagine the data measure hillslope and the saturation (water content) of the soil at each of the locations. Further imagine that we suspect the vegetation cover, bedrock and direction of the slope to vary across the study region. Therefore, whilst we could fit a regression line to explore the relationship between hillslope and saturation, we anticipate it would 'average away' local differences. In other words, we expect the relationship to vary across the study region (because of the local, unmeasured conditions).

One way to test for local variations in the regression relationship is to split the study region into parts, as in Figure 9.7(b), and fit a separate regression line to each. However, to do so encounters three problems. First, it assumes that where we divide the region coincides with where the relationship changes. Secondly, it implies the change is abrupt, occurring solely at the borders and not more gradual. Thirdly, the division of the study region into three parts is arbitrary.

A solution to the first problem is shown in Figure 9.7(c). Here we place a search window (called a kernel) around each point in turn, identifying all other points that fall within it. A regression line is then fitted to that subset of the data before moving sequentially from one point to the next across the study region, fitting a new regression line each time. We therefore obtain as many separate regression lines as there are points in the study region (or more, or less, if we choose to make the fitting points different from where the data are collected).

Because the search windows overlap, they imply gradual change across the study region. However, the search windows do not entirely allow for that gradualness. If points are either within the search window or not then the change is abrupt at its boundary. To address this, points are weighted so that those closest to the centre of the search window

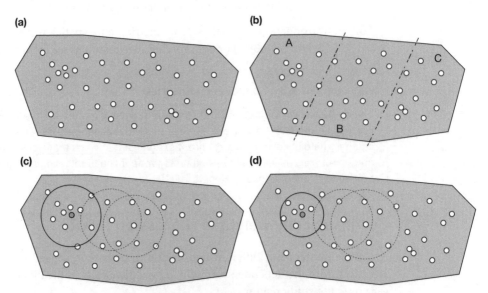

Figure 9.7 To show how regression relationships vary across a study region (a), separate regressions could be fitted to separate parts of the model (b). However, the division into parts is arbitrary. GWR works by having a moving window go from one observation to the next within the study region, each time fitting a regression model giving greatest weight to the observations with locations nearest to the centre. The window could be of a fixed size, as in (c), or adapt to the local density of observations, as in (d)

are given most weight in the regression relationship whilst points further away are given less weight, reducing to zero once the boundary of the search window is crossed. This is how the name **geographically weighted regression** (GWR) arises.

To avoid the criticism that the search window is of an arbitrary size, it is calibrated by repeatedly resizing it until it best fits the data. It is this process of calibration that is computationally demanding and makes the application of GWR difficult for very large data sets (Harris *et al.* 2010). In some cases it may be possible to circumnavigate the problem by specifying the size in advance based on theory or on previous studies.

In Figure 9.7(c) the search window has a fixed radius. However, the density of the points varies across the study region, so in some locations the window contains more points than in others. A solution is to allow the window to adapt its size to the local density. In Figure 9.7(d) the window shrinks or enlarges to contain six points always (plus the one at the centre). Six is rather few and the figure is illustrative only.

In summary, GWR fits weighted regression models sequentially across the study region. If (u_1, v_1) denotes the (easting, northing) or (longitude, latitude) location of the first data point, (u_2, v_2) the location of the second, etc., then estimates of the regression coefficients are made at each location:

$$\hat{y}_{(u_1, v_1)} = \beta_{0(u_1, v_1)} + \beta_{1(u_1, v_1)} x_1 + \beta_{2(u_1, v_1)} x_2 + \ldots$$

$$\hat{y}_{(u_2, v_2)} = \beta_{0(u_2, v_2)} + \beta_{1(u_2, v_2)} x_1 + \beta_{2(u_2, v_2)} x_2 + \ldots \quad (9.3)$$

$$\hat{y}_{(u_3, v_3)} = \beta_{0(u_3, v_3)} + \beta_{1(u_3, v_3)} x_1 + \beta_{2(u_3, v_3)} x_2 + \ldots$$

The final stage is to look at whether the (localised) beta values vary much across the study region. If they do, it means the regression relationship is not constant but varies geographically. This is useful information. Detecting and examining geographical differences is a helpful precursor to explaining why those geographical differences exist.

Table 9.4 summarises the results of using GWR to model the rates of non-participation in higher education in London. We have used an adaptive kernel that contains about 29 observations ($p = 0.0062$ of the data). The low number implies localised variations.

Looking at the effect of how many adults lack an educational qualification on the index of non-participation in the area, the table shows that whereas the standard (global) regression model associates a rise of one standardised unit on the predictor variable with an increase in the index of 0.640 units, actually the effect varies from $\beta = 0.156$ to $\beta = 0.977$ across the study region. Therefore, assuming all else is held constant, the effect is six times greater in some places compared with others. Even ignoring the most extreme cases and concentrating on the interquartile range, in some places the effect is 50% greater than in others ($\beta = 0.686$ vs $\beta = 0.463$).

A problem with this interpretation is all else is *not* held constant. If there is multicollinearity amongst the variables, their effects are interrelated; as one changes so do the others. Having indentified the geographical variation in the adult education variable, we refit the GWR model allowing it to vary across the study region but now holding all other variables constant. (We did this by fitting a partial regression model taking out the effects of the other variables on the index of non-participation and the adult education variable, then fitting a GWR model with the education variable as the sole predictor variable; see Chapter 7, Section 7.14.)

Table 9.4 Summarising the results of a GWR model estimating how the effects of the predictor variables on the index of non-participation vary across London

	Summary of the beta values					
	Min.	Q1	Median	Q3	Max.	Global
(Intercept)	−1.371	−0.248	−0.084	0.081	0.832	0.000
% adults in the area with no educational qualifications (standardised)	0.156	0.463	0.576	0.686	0.977	0.640
% households with four or more cars (standardised)	−1.168	−0.455	−0.295	−0.198	0.255	−0.276
% population Indian, Pakistani or Bangladeshi (standardised)	−1.887	−0.327	−0.232	−0.108	1.139	−0.263
Average attainment score in final year of compulsory education (standardised)	−0.464	−0.211	−0.134	−0.070	0.204	−0.137

Figure 9.8 maps the results of the new model, omitting places where the local effect of the educational variable is not significant at a 99% confidence level. What is left shows clear geographical patterning with especially an outer rim of places where the effect of no qualifications amongst the adult population is greatest upon whether 18–21 year olds attend university. There is also an area to the east of London where the variable appears to have no significance (at the 99% confidence level).

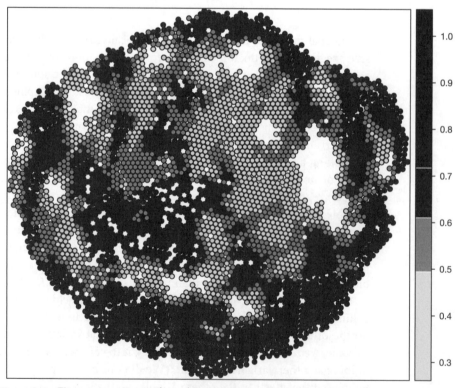

Figure 9.8 The varying effect of (prior) levels of educational attainment amongst adults on whether people aged 18–21 attend university (holding constant the effects of other variables). Areas where the effect is insignificant at a 99% confidence interval are omitted

> **Key concept 9.6 Geographically weighted regression (GWR)**
>
> **Geographically weighted regression** is a method to see if the relationships between the independent and the dependent variables vary across the study region.
>
> It fits a series of weighted regression models to overlapping parts of the study region, giving most weight to observations nearest to the centre of the search window. The search window is calibrated by a process of repeat testing to find the size that best fits the data.
>
> After the models have been fitted their beta and p values are compared to see if a variable has greater effect in one place more than others. In this way GWR reveals the sorts of local trends that a conventional regression model averages away.

GWR can be fitted using the spdep library for R, by purpose-built software available from http://ncg.nuim.ie/ncg/GWR/ and within ArcGIS 9.3. For a further introduction see Nakaya (2008). For a complete guide see Fotheringham *et al.* (2002).

9.8 Local indicator of spatial association (LISA)

GWR explores how the relationships between the dependent and independent variables vary across the study region. It detects local patterns of spatial association. However, it does contain a global parameter, one that is fixed for the entire study region. This is the size of the search window, held constant either by area or by the number of points contained within it. Calibrated for the entire study region, it is based on a global measure of spatial autocorrelation.

Moran's statistic, I, is also a global measure of spatial autocorrelation, essentially an average for the study region. Recall that the first stage in its calculation is to compare one observation with all its neighbours,

$$\sum_{j=1}^{n} w_{ij}\left(y_i - \bar{y}\right)\left(y_j - \bar{y}\right)$$

(cf. Equation (8.7)). The process is then repeated for all other observations.

Knowledge of this sequence leads to the local variant of Moran's I, introduced in Chapter 8. The Moran value for the index of non-participation in higher education across London is $I = 0.501$ but why create a single value for the entire region? Instead, we calculate a Moran's statistic specifically for the first observation and its neighbours, then another for the second observation and its neighbours, then the third, and so on across the study region. To do so gives us as many Moran's statistics as there are observations; each is a local statistic giving a measure of spatial association for a specific part of the study region. Comparing the **local Moran statistics** helps us to detect changing patterns of spatial association across the region.

Figure 9.9 shows the local Moran's statistics, excluding those that are not significant (locally) at a 95% confidence level. Note that some of the local Moran's values are high. The pattern of positive spatial autocorrelation that the global value detected appears driven by some quite concentrated patterns of non-participation. This suggests that residents might benefit from a targeted policy of supporting education in those places, including assistance for vocational learning.

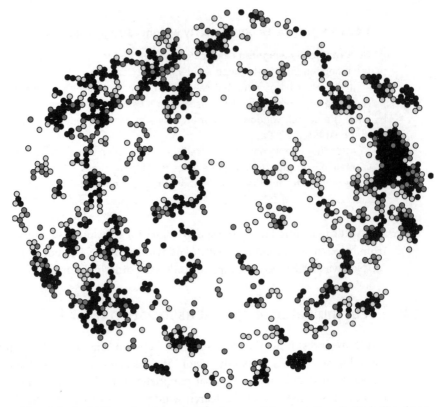

Figure 9.9 Local Moran's values showing how spatial associations in the index of non-participation vary across London. Insignificant values (at a 95% confidence interval) are omitted

The local Moran's statistic is an example of a **LISA: a local indicator of spatial association**. Anselin (1995) gives an operational definition of a LISA as any statistic that satisfies the following requirements. First, the LISA for each observation gives an indication of the extent of significant spatial clustering of similar values around that observation. Secondly, the sum of LISAs for all observations in the study region is proportional to the global indicator of spatial association.

> **Key concept 9.7 Local indicator of spatial association (LISA)**
>
> The **local Moran's statistic** separates the global Moran value into its constituent parts, allowing us to see if the pattern of spatial autocorrelation changes across the study region. It is an example of a **LISA**: a local indicator of spatial association.

9.9 Modelling at multiple scales

An important determinant of whether someone attends university (and particularly whether they do so at age 18–21) is their prior educational attainment, how well they did at school. Learning inequalities are related to social disadvantage and begin at an early age.

Equation 9.4 summarises a regression model predicting the attainment scores of pupils graduating from one of 200 randomly selected primary (elementary) schools in London in 2007. There are $n = 8212$ pupils in total.

$$y = 28.0(0.083) + 0.879(0.105)\text{older} + 0.237(0.092)\text{female} - 1.33(0.112)\text{ESL}$$
$$- 0.928(0.123)\text{black} + 0.657(0.137)\text{Asian} - 5.11(1.48)\text{traveller} + \varepsilon \qquad (9.4)$$

Each of the predictor variables is a dummy variable (cf. Chapter 7, Section 7.10) and is either true ($= 1$) or false ($= 0$) for each pupil (someone is either male or female, etc.).

The model says the expected attainment score for pupils is 28.0 but that value needs to be raised by 0.879 for those who are older (born in the first months of the academic year), by 0.237 for female pupils and by 0.657 for pupils recorded as being of an Asian ethnicity. It needs to be decreased by 1.33 for pupils for whom English is a second language (ESL), by 0.928 for pupils of a black ethnicity and by 5.11 for pupils recorded as travellers or gypsies. The predicted attainment score for a female pupil of black ethnicity born in October is therefore 28.2 ($= 28.0 + 0.879 + 0.237 - 0.928$). The predicted score for a white, male pupil born in February is simply 27.9.

The standard errors for the coefficients are shown in the brackets. Dividing the estimated beta value by its standard error gives a t value for which a p value can be obtained. Doing so, all of the predictor variables are found to be significant at a 99% confidence level, except the female variable, which is significant at a 95% confidence level.

For the purpose of the discussion we will presume that the various assumptions of regression have been checked and broadly met. That is not entirely likely given the absence of a continuous X variable, but we will let that problem pass. Instead we concentrate on another limitation of the model: it presumes that the only factors affecting attainment are ones measured as characteristics of the pupils themselves. Surely other factors are important, such as what type of school the pupils attend, its size and whether it attracts a mix of low- and high-attaining pupils?

To explore the effects of those school-level factors we may be tempted to fit the following using standard multivariate regression:

$$y = 30.8(0.277) + 0.895(0.103)\text{older} + 0.222(0.090)\text{female} - 0.943(0.111)\text{ESL}$$
$$- 0.548(0.122)\text{black} + 0.923(0.134)\text{Asian} - 4.28(1.44)\text{traveller}$$
$$- 0.164(0.009)\text{sch.var} - 0.009(0.012)\text{sch.size}$$
$$+ 0.427(0.169)\text{RC} + 0.489(0.173)\text{CoE} + \varepsilon \qquad (9.5)$$

The model now includes information about how much the attainment scores have varied within each school in the past (year.var), a measure of the school's size (sch.size), whether the pupil attends a Roman Catholic faith school (RC) and whether it is a Church of England school (CoE). All of the variables but school size appear significant at a 95% confidence level or greater.

Unfortunately, the interpretation of the results could be wrong. Consider, for example, the variable measuring how much the attainment scores have varied within the school. We have 8212 pupils in our data set, all of which attend a school. Hence, the variable contains 8212 observations. However, there are only 200 schools. Since the variable is measured at the school level, two or more pupils belonging to the same school must receive the same value; the values are duplicated. Is it therefore true to say we have $n = 8212$ observations for this and the other school-level variables or is the more truthful number closer to $n = 200$?

Getting the number right matters because it changes our uncertainty in the data. The less observations, the more uncertain we are; the standard errors rise, the t or z values fall, and variables we thought were significant at a given level of confidence may not be so.

A second problem with the model is that it implies all the errors are due to unexplained differences between the pupils (this is implied because a regression residual is calculated for each pupil but not for each school). This is inconsistent with our expectation of school effects. It is reasonable to expect differences between schools as well as between pupils.

Multilevel modelling provides the statistical framework to address these issues. It is used to model hierarchical data structures, in this example pupils within schools. It permits variables to be included for different levels of the model and for their effects to be examined simultaneously. We can separate pupil-level effects from school-level effects, and look at the unexplained errors at each level of the hierarchy.

Below are the results of the model, refitted using the multilevel software MLwiN:

$$y_{ij} = \beta_{0ij} + 0.918(0.100)\text{older}_{ij} + 0.226(0.088)\text{female}_{ij} - 0.879(0.113)\text{ESL}_{ij}$$
$$- 0.475(0.125)\text{black}_{ij} + 0.850(0.139)\text{Asian}_{ij} - 3.52(1.41)\text{traveller}_{ij}$$
$$- 0.166(0.016)\text{class.var}_{j} - 0.004(0.024)\text{sch.size}_{j}$$
$$+ 0.340(0.298)\text{RC}_{j} + 0.367(0.289)\text{CoE}_{j} \tag{9.6}$$
$$\beta_{0ij} = 30.8(0.497) + u_{0j} + \varepsilon_{0ij}$$
$$\left[u_{0j}\right]: \text{N}\left(0, \Omega_{u}\right) = \left[1.00(0.144)\right]$$
$$\left[\varepsilon_{0ij}\right]: \text{N}\left(0, \Omega_{\varepsilon}\right) = \left[15.5(0.245)\right]$$

To understand the results, begin by noting that there are two levels to the model: the pupil level, with subscript i, and the school level, with subscript j. All the pupil-level variables appear with both subscripts as a reminder that all pupils attended a primary school. These are distinguished from variables that describe the school, not the pupils, for which only the subscript j appears.

Comparing Equation 9.6 with 9.5, using the multilevel model has not greatly changed the estimate of each variable's effect on attainment. Nor do the standard errors (shown in brackets) for the pupil-level variables differ much. However, for the school-level variables the standard errors have been raised, recognising there are far fewer schools than there are pupils in them. As a rule of thumb, if the estimated effect divided by its standard error is of magnitude greater than two, it is significant at about a 95% confidence level. In this way we find that our previous model did indeed mislead: being a faith school is not the significant predictor of attainment we thought it was ($0.340/0.298 = 1.14$; $0.367/0.289 = 1.27$).

The next line of the model summary is allowing for random differences between pupils (ε_{0ij}) and between schools (u_{0j}). Modelling the differences between schools is like fitting a separate y intercept (a separate baseline effect) for each school and then seeing how those different intercepts vary around the intercept for all pupils and schools, which is β_{0ij}.

The estimate of that school-level variance is in the next line, with the assumption that it is drawn from a normal population (which is what the N indicates). Unsurprisingly, the pupil-level variance is much greater (differences between pupils exceed differences between schools: 15.5 vs 1.00) but there are still significant unexplained differences between schools (1.00/0.144 > 2).

There is no reason to restrict the model to two levels; Equation 9.7 has a third. The model now includes a measure of income poverty within the local educational authority (LEA) to which the school belongs. The hierarchy is now pupils (i) within schools (j) within LEAs (k). The new variable is of borderline significance, 1.519/0.866 = 1.75, but there is no significant residual variance at the LEA level (0.023/0.032 = 0.719). These results suggest the third level could be omitted.

$$y_{ij} = \beta_{0ijk} + 0.921(0.099)\text{older}_{ijk} + 0.227(0.087)\text{female}_{ijk} - 0.910(0.115)\text{ESL}_{ijk}$$
$$- 0.488(0.124)\text{black}_{ijk} + 0.833(0.140)\text{Asian}_{ijk} - 3.58(1.42)\text{traveller}_{ijk}$$
$$- 0.186(0.016)\text{class.var}_{jk} + 1.52(0.866)\text{LEA.poverty}_{k}$$

$$\beta_{0ij} = 30.4(0.452) + v_{0k} + u_{0jk} + \varepsilon_{0ijk}$$

$$\left[v_{0k}\right] : N\left(0, \Omega_u\right) = \left[0.023(0.032)\right]$$

$$\left[u_{0jk}\right] : N\left(0, \Omega_u\right) = \left[0.984(0.144)\right]$$

$$\left[\varepsilon_{0ijk}\right] : N\left(0, \Omega_\varepsilon\right) = \left[15.5(0.246)\right]$$

(9.7)

As here, multilevel modelling can be used to model at multiple scales simultaneously and to explore how individual behaviours and characteristics are shaped by the places in which the pupils live or by the organisations they attend. The models can be extended to explore whether the effect of a predictor variable varies at different levels of the hierarchy. For example, the effect on attainment of speaking English as a second language is significantly greater in some schools than in others. Multilevel models can also be used to model longitudinally, to consider changes over time (for example, have schools and/or LEAs become more socially segregated in the UK since the introduction of more active competition and choice?) (Goldstein and Noden 2003).

Because multilevel models can consider people in places they are sometimes used to generate evidence of a **neighbourhood effect**. This is an effect that makes a measurable difference to individuals, their behaviours, attitudes and/or lifestyles but is not reducible to individual choices and characteristics alone. Its existence is posited on the view that individuals do not live in a social vacuum but interact within their local community and environments such that where people live, and who and what they meet, have an effect on what they do or become. Examples of a neighbourhood effect include (unconscious?) pressures to conform, be it an increased likelihood to drive a prestigious car because neighbours do, the influence of good or bad role models, or the effect of tradition and social upbringing on voting behaviour. Neighbourhood effects are difficult to measure but multilevel modelling can provide strong circumstantial evidence for them (Harris *et al.* 2007).

A comprehensive introduction to multilevel modelling with accompanying software and tutorials is available at The Centre for Multilevel Modelling (www.cmm.bristol.ac.uk) (see also Goldstein 2010; Hox 2010; Luke 2004; Snijders and Bosker 1999).

> **Key concept 9.8 Multilevel modelling**
>
> **Multilevel modelling** is used to model and to separate out the effects of various predictor variables on a dependent variable at different levels of hierarchy. The hierarchy could be geographical (for example, people in places in neighbourhood types), longitudinal or a mixture of the two, amongst other possibilities.

9.10 Geography, computation and statistics

The development of spatial regression and multilevel models has been made possible by advances in computation. Whereas conventional regression models use ordinary least squares (OLS) to fit the regression line, more complex data structures – including those that consider geography and geographical hierarchies – require more complex approaches. In broad terms these tend to be iterative, using the computer to try out possibilities and search for a solution. They also learn from the data, using them to determine the most probable answers. Techniques like GWR are characterised by repeat fitting: they go across the study region looking for localised patterns of association. These are especially demanding computationally. Even in this day and age, analyses of large data sets could take hours, days or even weeks to run.

A helpful development is the ability to explore data 'on the fly', especially using graphical techniques to enhance understanding of the data. Software such as GeoDa incorporate dynamic linking, which means you could click on an outlier in a bar chart and see that observation highlighted on a map. This aids understanding of the data, knowing which are the more unusual observations and where they are located.

The coming together of the statistical, the graphical and the visual to explore, interact with and identify trends within data is sometimes described as **scientific visualisation**. Here the graphics are more than a way of just presenting the data or summarising the results of an analysis. They become instead an intrinsic part of the scientific investigation, of the process by which ideas and knowledge are shaped and hypotheses formed and tested.

GIS software have provided a visual interface to geographical data sets using the medium of the map. New developments in open source software such as R allow GIS files to be read into the software environment, a weights matrix to be generated automatically, a spatial regression model to be fitted and the results plotted using the high-quality graphics available to the user. There is increasing integration between geographical information science, computer science and statistics.

From Chapter 7 to here the move has been towards thinking geographically and revealing geographical differences. Whereas the assumptions of standard regression make geography a nuisance, the spatially lagged *y* model sees it as a consequence of an underlying process, GWR looks at how relationships vary across a study region and the use of multilevel modelling posits context matters. Harnessing the power of computers, it is possible to focus more and more on the detail and away from a one-size-fits-all way of modelling.

Given this, we should remind ourselves that an overly complex model that cannot be explained and lacking any theoretical justification is not helpful. We should also remember that it is not only differences between people and places that matter, but also their commonalities and similarities too. The methods of modelling we have looked at in this chapter assist in getting the balance right: they do not discard geographical information or act blindly to geographical patterns of association, but instead ground them within a statistical framework that allows for differences due to chance to be detected, the effects of geographical relationships to be considered, and to help determine what causes what – *and where.*

Key points

- Geographical data are records of what happened at some location on or near the Earth's surface and of where it happened.

- Spatial analysis is a set of methods that incorporate both the attribute and the locational information from geographical data.

- Conventional regression analysis is not a geographical method of analysis but can be extended to incorporate a (spatial) weights matrix defining how parts of a map are related.

- These spatial regression models include the spatial error model and the spatially lagged *y* model and are drawn from the spatial econometrics literature.

- Cartograms are a useful way of presenting data in a way that gives increased emphasis to places with the highest amount of a variable of interest whilst still retaining a sense of where places are located in relation to each other.

- Moran's *I* statistic is a measure of spatial autocorrelation, averaged across a study region. Local Moran's statistic is an example of a LISA: a local indicator of spatial association. The LISA for each observation gives an indication of the extent of significant spatial clustering of similar values around that observation.

- Geographically weighted regression (GWR) is used to look at how the relationships between the dependent and independent variables vary across the study region. It reveals the sorts of localised patterns and deviations from the norm that are concealed or averaged away by standard one-size-fits-all regression.

- Multilevel models are used for modelling hierarchies, for example pupils within schools with local education authorities. They can be used to model at multiple scales simultaneously, allowing the effects of the predictor variables to be explored at difference scales and providing robust estimates of their standard errors when doing so.

- Developments in computing have seen increased integration between geographical information, visualisation and statistical software.

- Although it is important to recognise geographical differences, where they exist and why, sometimes those differences are due to chance or coexist with a large degree of similarity between the objects or observations being modelled. Adopting a spatial and statistical framework for analysis helps to explore these possibilities.

References

Anselin, L. (1995) Local indicators of spatial association – LISA. *Geographical Analysis,* 27, 93–115.

Bivand, R.S., Pebesma, E.J. and Gomez-Rubio, V. (2008) *Applied Spatial Data Analysis with R,* New York: Springer.

Dorling, D. (1996) *Area Cartograms: Their Use and Creation,* Norwich: Environmental Publications.

Fotheringham, A.S., Brunsdon, C. and Charlton, M. (2002) *Geographically Weighted Regression: The Analysis of Spatially Varying Relationships,* Chichester: Wiley.

Goldstein, H. (2010) *Multilevel Statistical Models,* 4th edn, Chichester: Wiley-Blackwell.

Goldstein, H. and Noden, P. (2003) Modelling social segregation. *Oxford Review of Education,* 29(2), 225–237.

Harris, R., Johnston, R. and Burgess, S. (2007) Neighborhoods, ethnicity and school choice: developing a statistical framework for geodemographic analysis. *Population Research and Policy Review,* 26(5), 553–579.

Harris, R., Singleton, A., Grose, D., Brunsdon, C. and Longley, P. (2010) Grid-enabling geographically weighted regression: a case study of participation in higher education in England. *Transactions in GIS,* 14(1), 43–61.

Hepple, L.W. (1998) Exact testing for spatial autocorrelation among regression residuals. *Environment and Planning A,* 30(1), 85–108.

Hepple, L.W. (2001) Multiple regression and spatial policy analysis: George Udny Yule and the origins of statistical social science. *Environment and Planning D: Society and Space,* 19(4), 385–407.

Hox, J. (2010) *Multilevel Analysis,* 2nd edn, Abingdon: Routledge Academic.

LeSage, J. and Pace, R.K. (2008) *Introduction to Spatial Econometrics,* Boca Raton, FL: Chapman and Hall/CRC Press.

Longley, P.A., Goodchild, M., Maguire, D.J. and Rhind, D.W. (2010) *Geographic Information Systems and Science,* 3rd edn, Chichester: Wiley.

Luke, D.D.A. (2004) *Multilevel Modeling,* Thousand Oaks, CA: Sage.

Monmonier, M. (2010) *No Dig, No Fly, No Go: How Maps Restrict and Control,* Chicago: Chicago University Press.

Moran, P. (1948) The interpretation of statistical maps. *Journal of the Royal Statistical Society Series B,* 10, 245–251.

Moran, P. (1950) Notes on continuous stochastic phenomena. *Biometrika,* 37, 17–23.

Nakaya, T. (2008) Geographically weighted regression (GWR). In *Encyclopedia of Geographical Information Science,* London: Sage, pp.179–184.

Pace, R.K. and LeSage, J. (2010) Spatial econometrics. In *Handbook of Spatial Statistics,* Boca Raton, FL: Chapman and Hall/CRC Press, pp.245–262.

Snijders, P.T.A. and Bosker, P.R. (1999) *Multilevel Analysis: An Introduction to Basic and Advanced Multilevel Modeling,* London: Sage.

Thomas, B. and Dorling, D. (2007) *Identity in Britain: A Cradle-to-grave Atlas,* Bristol: Policy Press.

Tiefelsdorf, M. (1998) Some practical applications of Moran's I's exact conditional distribution. *Papers in Regional Science,* 77, 101–129.

Ward, M.D. and Gleditsch, P.K.S. (2008) *Spatial Regression Models,* Thousand Oaks, CA: Sage.

Epilogue

As we reach the close of this book it has been a long journey from descriptive statistics, through research design and inferential statistics to explanatory statistics, and from non-geographical to explicitly spatial methods of analysis.

There is a lot more we could say but this book has plenty of words already. Let us instead provide a summary of what has been learned and suggest some reading to take your knowledge and understanding further, in ways of your choosing.

Chapter 1 made the case for knowing about statistics as a transferable skill and to be equipped for social and political debate. For a provocative use of regression analysis to make the case against social inequality see Wilkinson and Pickett (2009) (and for a counterview, Snowdon 2010). To learn how statistics can be manipulated, massaged and misrepresented to promote powerful interests read Best (2001).

Chapter 2 was about using descriptive statistics and simple graphical techniques to explore and make sense of data. They can also be used for deliberate distortion of the data and to mislead. To learn how (not) to do so, the classic texts are Huff (1954) and Monmonier (1996). For an outstanding text on the display of quantitative information see Tufte (2001).

Chapter 3 talked about the normal curve, the properties of which then provided the basis for inferential statistics in Chapter 5. It is interesting and informative to know something of the history of statistics and how ideas and thinking evolved. A very readable book for doing so is Salsburg (2002).

Chapter 4 was about the principles of research design and effective data collection. There are many books on this subject including Creswell (2008).

Chapter 6 described the role of hypothesis testing, looking at whether a result could be described as statistically significant. We looked at issues of statistical power and the importance of considering whether a result is of theoretical/substantive significance. For a book that warns against a preoccupation with statistical significance alone, see Ziliak and McCloskey (2008).

Chapter 7 was about regression analysis. To learn more about statistical modelling and a defence of it for causal inference read Freedman (2009a; 2009b).

Chapter 8 moved to modelling point patterns, 'hotspot analysis' and ways of measuring patterns of spatial autocorrelation in data. To read more try Lloyd (2009), McKillup and Dyar (2010), O'Sullivan and Unwin (2010) and Fotheringham *et al.* (2000).

Finally, Chapter 9 looked at spatial regression models, geographically weighted regression and multilevel modelling. Introductions to these include Fotheringham *et al.* (2002), Ward and Gleditsch (2008) and the tutorials available at www.cmm .bristol.ac.uk.

Unsurprisingly, these suggestions are only the tip of a very large iceberg. Areas we have not covered in any detail include geographical information science (Wilson and Fotheringham 2007), area-based classifications (Harris *et al.* 2005), process modelling using artificial intelligence such as cellular automata, agent-based models and fractals

(Batty 2007), the application of statistics to so-called neogeography (Turner 2006) or the search for 'shared standards' and mutual learning in quantitative and qualitative research (Brady 2004; Oakley 2000).

Clearly there is much more that could be taught and learned. For now we hope to have given you an overview of statistical concepts and ideas that equip you to take your research interests in the direction of your choosing. To go back to where we started, the preface, we hope that what we have done is given you wheels to travel.

References

Batty, M. (2007) *Cities and Complexity: Understanding Cities with Cellular Automata, Agent-Based Models, and Fractals,* Reading, MA: MIT Press.

Best, J. (2001) *Damned Lies and Statistics: Untangling Numbers from the Media, Politicians and Activists,* Berkeley, CA: University of California Press.

Brady, H.E. (2004) *Rethinking Social Inquiry: Diverse Tools, Shared Standards,* Lanham, MD: Rowman & Littlefield.

Creswell, J.W. (2008) *Research Design: Qualitative, Quantitative, and Mixed Methods Approaches,* 3rd edn, Thousand Oaks, CA: Sage.

Fotheringham, A.S., Brunsdon, C. and Charlton, M. (2002) *Geographically Weighted Regression: The Analysis of Spatially Varying Relationships,* Chichester: Wiley.

Fotheringham, P.A.S., Brunsdon, C. and Charlton, M. (2000) *Quantitative Geography: Perspectives on Spatial Data Analysis,* London: Sage.

Freedman, D.A. (2009a) *Statistical Models and Causal Inference: A Dialogue with the Social Sciences,* Cambridge: Cambridge University Press.

Freedman, D.A. (2009b) *Statistical Models: Theory and Practice,* 2nd edn, Cambridge: Cambridge University Press.

Harris, R., Sleight, P. and Webber, R. (2005) *Geodemographics: GIS and Neighbourhood Targeting,* Chichester: Wiley.

Huff, D. (1954) *How to Lie With Statistics,* New York: W W Norton.

Lloyd, C. (2009) *Spatial Data Analysis: An Introduction for GIS Users,* Oxford: Oxford University Press.

McKillup, S. and Dyar, M.D. (2010) *Geostatistics Explained: An Introductory Guide for Earth Scientists,* Cambridge: Cambridge University Press.

Monmonier, M.S. (1996) *How to Lie with Maps,* 2nd edn, Chicago: Chicago University Press.

Oakley, A. (2000) *Experiments in Knowing: Gender and Method in the Social Sciences,* Cambridge: Polity Press.

O'Sullivan, D. and Unwin, D. (2010) *Geographic Information Analysis,* 2nd edn, Chichester: Wiley.

Salsburg, D. (2002) *The Lady Tasting Tea: How Statistics Revolutionized Science in the Twentieth Century,* 2nd edn, New York: Owl Books.

Snowdon, C.J. (2010) *The Spirit Level Delusion: Fact-checking the Left's New Theory of Everything,* Democracy Institute/Little Dice.

Tufte, E.R. (2001) *The Visual Display of Quantitative Information,* 2nd edn, Cheshire, CT: Graphics Press.

Turner, A. (2006) *Introduction to Neogeography,* Sebastopol, CA: O'Reilly.

Ward, M.D. and Gleditsch, P.K.S. (2008) *Spatial Regression Models,* Thousand Oaks, CA: Sage.

Wilkinson, R. and Pickett, K. (2009) *The Spirit Level: Why More Equal Societies Almost Always Do Better,* London: Allen Lane.

Wilson, J.P. and Fotheringham, A.S. (2007) *Handbook of Geographic Information Science,* Chichester: Wiley-Blackwell.

Ziliak, S.T. and McCloskey, D.N. (2008) *The Cult of Statistical Significance: How the Standard Error Costs Us Jobs, Justice, and Lives,* Ann Arbor, MI: University of Michigan Press.

Appendix

The area under a standard normal curve to the given values of *z*
(See Chapter 3, Section 3.8 for a discussion of it)

z	0.00	0.01	0.02	0.03	0.04	0.05	0.06	0.07	0.08	0.09
-3.9	0.0000	0.0000	0.0000	0.0000	0.0000	0.0000	0.0000	0.0000	0.0000	0.0000
-3.8	0.0001	0.0001	0.0001	0.0001	0.0001	0.0001	0.0001	0.0001	0.0001	0.0001
-3.7	0.0001	0.0001	0.0001	0.0001	0.0001	0.0001	0.0001	0.0001	0.0001	0.0001
-3.6	0.0002	0.0002	0.0001	0.0001	0.0001	0.0001	0.0001	0.0001	0.0001	0.0001
-3.5	0.0002	0.0002	0.0002	0.0002	0.0002	0.0002	0.0002	0.0002	0.0002	0.0002
-3.4	0.0003	0.0003	0.0003	0.0003	0.0003	0.0003	0.0003	0.0003	0.0003	0.0002
-3.3	0.0005	0.0005	0.0005	0.0004	0.0004	0.0004	0.0004	0.0004	0.0004	0.0003
-3.2	0.0007	0.0007	0.0006	0.0006	0.0006	0.0006	0.0006	0.0005	0.0005	0.0005
-3.1	0.0010	0.0009	0.0009	0.0009	0.0008	0.0008	0.0008	0.0008	0.0007	0.0007
-3.0	0.0013	0.0013	0.0013	0.0012	0.0012	0.0011	0.0011	0.0011	0.0010	0.0010
-2.9	0.0019	0.0018	0.0018	0.0017	0.0016	0.0016	0.0015	0.0015	0.0014	0.0014
-2.8	0.0026	0.0025	0.0024	0.0023	0.0023	0.0022	0.0021	0.0021	0.0020	0.0019
-2.7	0.0035	0.0034	0.0033	0.0032	0.0031	0.0030	0.0029	0.0028	0.0027	0.0026
-2.6	0.0047	0.0045	0.0044	0.0043	0.0041	0.0040	0.0039	0.0038	0.0037	0.0036
-2.5	0.0062	0.0060	0.0059	0.0057	0.0055	0.0054	0.0052	0.0051	0.0049	0.0048
-2.4	0.0082	0.0080	0.0078	0.0075	0.0073	0.0071	0.0069	0.0068	0.0066	0.0064
-2.3	0.0107	0.0104	0.0102	0.0099	0.0096	0.0094	0.0091	0.0089	0.0087	0.0084
-2.2	0.0139	0.0136	0.0132	0.0129	0.0125	0.0122	0.0119	0.0116	0.0113	0.0110
-2.1	0.0179	0.0174	0.0170	0.0166	0.0162	0.0158	0.0154	0.0150	0.0146	0.0143
-2.0	0.0228	0.0222	0.0217	0.0212	0.0207	0.0202	0.0197	0.0192	0.0188	0.0183
-1.9	0.0287	0.0281	0.0274	0.0268	0.0262	0.0256	0.0250	0.0244	0.0239	0.0233
-1.8	0.0359	0.0351	0.0344	0.0336	0.0329	0.0322	0.0314	0.0307	0.0301	0.0294
-1.7	0.0446	0.0436	0.0427	0.0418	0.0409	0.0401	0.0392	0.0384	0.0375	0.0367
-1.6	0.0548	0.0537	0.0526	0.0516	0.0505	0.0495	0.0485	0.0475	0.0465	0.0455
-1.5	0.0668	0.0655	0.0643	0.0630	0.0618	0.0606	0.0594	0.0582	0.0571	0.0559
-1.4	0.0808	0.0793	0.0778	0.0764	0.0749	0.0735	0.0721	0.0708	0.0694	0.0681
-1.3	0.0968	0.0951	0.0934	0.0918	0.0901	0.0885	0.0869	0.0853	0.0838	0.0823
-1.2	0.1151	0.1131	0.1112	0.1093	0.1075	0.1056	0.1038	0.1020	0.1003	0.0985
-1.1	0.1357	0.1335	0.1314	0.1292	0.1271	0.1251	0.1230	0.1210	0.1190	0.1170
-1.0	0.1587	0.1562	0.1539	0.1515	0.1492	0.1469	0.1446	0.1423	0.1401	0.1379
-0.9	0.1841	0.1814	0.1788	0.1762	0.1736	0.1711	0.1685	0.1660	0.1635	0.1611
-0.8	0.2119	0.2090	0.2061	0.2033	0.2005	0.1977	0.1949	0.1922	0.1894	0.1867
-0.7	0.2420	0.2389	0.2358	0.2327	0.2296	0.2266	0.2236	0.2206	0.2177	0.2148
-0.6	0.2743	0.2709	0.2676	0.2643	0.2611	0.2578	0.2546	0.2514	0.2483	0.2451
-0.5	0.3085	0.3050	0.3015	0.2981	0.2946	0.2912	0.2877	0.2843	0.2810	0.2776
-0.4	0.3446	0.3409	0.3372	0.3336	0.3300	0.3264	0.3228	0.3192	0.3156	0.3121
-0.3	0.3821	0.3783	0.3745	0.3707	0.3669	0.3632	0.3594	0.3557	0.3520	0.3483
-0.2	0.4207	0.4168	0.4129	0.4090	0.4052	0.4013	0.3974	0.3936	0.3897	0.3859
-0.1	0.4602	0.4562	0.4522	0.4483	0.4443	0.4404	0.4364	0.4325	0.4286	0.4247
0.0	0.5000	0.5040	0.5080	0.5120	0.5160	0.5199	0.5239	0.5279	0.5319	0.5359
0.1	0.5398	0.5438	0.5478	0.5517	0.5557	0.5596	0.5636	0.5675	0.5714	0.5753
0.2	0.5793	0.5832	0.5871	0.5910	0.5948	0.5987	0.6026	0.6064	0.6103	0.6141
0.3	0.6179	0.6217	0.6255	0.6293	0.6331	0.6368	0.6406	0.6443	0.6480	0.6517
0.4	0.6554	0.6591	0.6628	0.6664	0.6700	0.6736	0.6772	0.6808	0.6844	0.6879

continued

z	0.00	0.01	0.02	0.03	0.04	0.05	0.06	0.07	0.08	0.09
0.5	0.6915	0.6950	0.6985	0.7019	0.7054	0.7088	0.7123	0.7157	0.7190	0.7224
0.6	0.7257	0.7291	0.7324	0.7357	0.7389	0.7422	0.7454	0.7486	0.7517	0.7549
0.7	0.7580	0.7611	0.7642	0.7673	0.7704	0.7734	0.7764	0.7794	0.7823	0.7852
0.8	0.7881	0.7910	0.7939	0.7967	0.7995	0.8023	0.8051	0.8078	0.8106	0.8133
0.9	0.8159	0.8186	0.8212	0.8238	0.8264	0.8289	0.8315	0.8340	0.8365	0.8389
1.0	0.8413	0.8438	0.8461	0.8485	0.8508	0.8531	0.8554	0.8577	0.8599	0.8621
1.1	0.8643	0.8665	0.8686	0.8708	0.8729	0.8749	0.8770	0.8790	0.8810	0.8830
1.2	0.8849	0.8869	0.8888	0.8907	0.8925	0.8944	0.8962	0.8980	0.8997	0.9015
1.3	0.9032	0.9049	0.9066	0.9082	0.9099	0.9115	0.9131	0.9147	0.9162	0.9177
1.4	0.9192	0.9207	0.9222	0.9236	0.9251	0.9265	0.9279	0.9292	0.9306	0.9319
1.5	0.9332	0.9345	0.9357	0.9370	0.9382	0.9394	0.9406	0.9418	0.9429	0.9441
1.6	0.9452	0.9463	0.9474	0.9484	0.9495	0.9505	0.9515	0.9525	0.9535	0.9545
1.7	0.9554	0.9564	0.9573	0.9582	0.9591	0.9599	0.9608	0.9616	0.9625	0.9633
1.8	0.9641	0.9649	0.9656	0.9664	0.9671	0.9678	0.9686	0.9693	0.9699	0.9706
1.9	0.9713	0.9719	0.9726	0.9732	0.9738	0.9744	0.9750	0.9756	0.9761	0.9767
2.0	0.9772	0.9778	0.9783	0.9788	0.9793	0.9798	0.9803	0.9808	0.9812	0.9817
2.1	0.9821	0.9826	0.9830	0.9834	0.9838	0.9842	0.9846	0.9850	0.9854	0.9857
2.2	0.9861	0.9864	0.9868	0.9871	0.9875	0.9878	0.9881	0.9884	0.9887	0.9890
2.3	0.9893	0.9896	0.9898	0.9901	0.9904	0.9906	0.9909	0.9911	0.9913	0.9916
2.4	0.9918	0.9920	0.9922	0.9925	0.9927	0.9929	0.9931	0.9932	0.9934	0.9936
2.5	0.9938	0.9940	0.9941	0.9943	0.9945	0.9946	0.9948	0.9949	0.9951	0.9952
2.6	0.9953	0.9955	0.9956	0.9957	0.9959	0.9960	0.9961	0.9962	0.9963	0.9964
2.7	0.9965	0.9966	0.9967	0.9968	0.9969	0.9970	0.9971	0.9972	0.9973	0.9974
2.8	0.9974	0.9975	0.9976	0.9977	0.9977	0.9978	0.9979	0.9979	0.9980	0.9981
2.9	0.9981	0.9982	0.9982	0.9983	0.9984	0.9984	0.9985	0.9985	0.9986	0.9986
3.0	0.9987	0.9987	0.9987	0.9988	0.9988	0.9989	0.9989	0.9989	0.9990	0.9990
3.1	0.9990	0.9991	0.9991	0.9991	0.9992	0.9992	0.9992	0.9992	0.9993	0.9993
3.2	0.9993	0.9993	0.9994	0.9994	0.9994	0.9994	0.9994	0.9995	0.9995	0.9995
3.3	0.9995	0.9995	0.9995	0.9996	0.9996	0.9996	0.9996	0.9996	0.9996	0.9997
3.4	0.9997	0.9997	0.9997	0.9997	0.9997	0.9997	0.9997	0.9997	0.9997	0.9998
3.5	0.9998	0.9998	0.9998	0.9998	0.9998	0.9998	0.9998	0.9998	0.9998	0.9998
3.6	0.9998	0.9998	0.9999	0.9999	0.9999	0.9999	0.9999	0.9999	0.9999	0.9999
3.7	0.9999	0.9999	0.9999	0.9999	0.9999	0.9999	0.9999	0.9999	0.9999	0.9999
3.8	0.9999	0.9999	0.9999	0.9999	0.9999	0.9999	0.9999	0.9999	0.9999	0.9999
3.9	1.0000	1.0000	1.0000	1.0000	1.0000	1.0000	1.0000	1.0000	1.0000	1.0000

The value of *t* required to give a 90%, 95%, 99% and 99.9% confidence interval for the given degrees of freedom

(See Chapter 5, Section 5.11 for discussion of it)

df	(90%) p = 0.05	(95%) p = 0.025	(99%) p = 0.005	(99.9%) p = 0.0005
1	−6.314	−12.706	−63.657	−636.619
2	−2.920	−4.303	−9.925	−31.599
3	−2.353	−3.182	−5.841	−12.924
4	−2.132	−2.776	−4.604	−8.610
5	−2.015	−2.571	−4.032	−6.869
6	−1.943	−2.447	−3.707	−5.959
7	−1.895	−2.365	−3.499	−5.408
8	−1.860	−2.306	−3.355	−5.041
9	−1.833	−2.262	−3.250	−4.781
10	−1.812	−2.228	−3.169	−4.587
11	−1.796	−2.201	−3.106	−4.437
12	−1.782	−2.179	−3.055	−4.318
13	−1.771	−2.160	−3.012	−4.221
14	−1.761	−2.145	−2.977	−4.140
15	−1.753	−2.131	−2.947	−4.073
16	−1.746	−2.120	−2.921	−4.015
17	−1.740	−2.110	−2.898	−3.965
18	−1.734	−2.101	−2.878	−3.922
19	−1.729	−2.093	−2.861	−3.883
20	−1.725	−2.086	−2.845	−3.850
21	−1.721	−2.080	−2.831	−3.819
22	−1.717	−2.074	−2.819	−3.792
23	−1.714	−2.069	−2.807	−3.768
24	−1.711	−2.064	−2.797	−3.745
25	−1.708	−2.060	−2.787	−3.725
26	−1.706	−2.056	−2.779	−3.707
27	−1.703	−2.052	−2.771	−3.690
28	−1.701	−2.048	−2.763	−3.674
29	−1.699	−2.045	−2.756	−3.659
30	−1.697	−2.042	−2.750	−3.646

Index